THE PRINCIPLES AND APPLICATIONS OF VARIATIONAL METHODS

MARTIN BECKER

RESEARCH MONOGRAPH NO. 27
THE M.I.T. PRESS, CAMBRIDGE, MASSACHUSETTS

Second Printing, October 1970

ISBN 0 262 02009 2   (hardcover)

Library of Congress Catalog Card Number: 64-25215

Printed in the United States of America

FOREWORD

There has long been a need in science and engineering for systematic publication of research studies larger in scope than a journal article but less ambitious than a finished book. Much valuable work of this kind is now published only in a semiprivate way, perhaps as a laboratory report, and so may not find its proper place in the literature of the field. The present contribution is the twenty-seventh of the M.I.T. Press Research Monographs, which we hope will make selected timely and important research studies readily accessible to libraries and to the independent worker.

J. A. Stratton

PREFACE

The advice and assistance of Professor Henri Fenech, Massachusetts Institute of Technology, was essential to the early and successful completion of this work. His contagious enthusiasm for variational methods was a major factor governing the author's decision to work in this area.

The author would like to acknowledge several stimulating and profitable discussions with Drs. D. E. Dougherty, N. C. Francis, and D. S. Selengut. Dr. I. Wall provided valuable assistance in the use of depletion programs. Professors E. P. Gyftopoulos and K. F. Hansen read an earlier draft of the thesis. Their recommendations concerning the organization of the final manuscript were very useful. Professor M. Benedict read the final draft when Professor Fenech unfortunately became too ill to do so. Dr. Benedict's assistance in this matter is greatly appreciated.

The financial support of the Atomic Energy Commission and the National Science Foundation throughout the author's stay at M. I. T. is greatly appreciated.

The depletion studies were made using the facilities of the M. I. T. Computation Center.

The author feels most indebted to his wife Jean, whose typing of the first draft was the least of her contributions.

The author also wishes to acknowledge the typing efforts of Miss Rachel Sprinsky, who worked long hours so that the manuscript could be completed on time.

Cambridge, Massachusetts
July 1, 1964   Martin Becker

CONTENTS

Contents

Chapter 1

INTRODUCTION

Most practical problems in science and engineering generally
are described by complicated equations that cannot be solved exactly by current analytical methods. When faced with such complex problems, we usually must resort to methods that yield approximate solutions.

In some cases, simplification of the physical model may lead
to simpler equations. In other cases, the equations may be similar in form to equations whose solutions are known, and perturbation theory can be applied. A third approach is a direct attack
on the equations, with finite differences replacing derivatives, and
with discrete ordinates replacing some or all of the continuous
variables. There are, in addition, many other methods of approximation, such as series solution, iterative procedures, and
so forth.

This work is concerned with methods based on the calculus of
variations, which seek to combine "trial functions" (guesses about
the form of the solution) into a satisfactory approximate solution.
These methods are, of course, most easily applied in cases where
the analyst has good insight (based on intuition, experience, and
so forth) into the form of the solution. Variational methods consist essentially of finding a functional whose variation yields the
equation of interest as its "Euler equation," substituting the trial
solution into the functional, and, by taking variations with respect
to some adjustable parameters, determining these parameters and
thereby the "best" approximate solution of the equation (with the
given trial functions). Present variational methods have some serious disadvantages that interfere with the application of variational methods to certain types of problems; these limitations fall
into three categories:

   1. The lack of flexibility of "self-sufficient" methods (those
that do not require the introduction of adjoint functions).

   2. The additional complexity introduced by adjoint functions
in non-self-sufficient problems.

   3. The possibility that positive and negative errors may tend
to cancel each other, and, consequently, the possibility that an
accurate functional may be accompanied by an inaccurate function.

1

The main object of this work is to develop improved variational methods that do not have these limitations and are applicable to general types of problems.

After the basic principles of variational and other trial function methods are reviewed in Chapter 2, Chapter 3 undertakes to propose such new techniques. The viewpoint advanced is that a fundamental rule of variational methods, which states that the functional should yield the equation of interest as its Euler equation, is unnecessarily restrictive. It is deemed sufficient that the functional have an Euler equation which has the solution of the equation of interest as a particular solution. It is shown that of these functionals, those of the least-squares type best satisfy a set of criteria put forth to measure the merits of functionals in general problems.

An analysis is made of least-squares variational methods. It is shown how "extraneous" boundary conditions introduced by the method can be used profitably to improve the accuracy of the results with little increase in effort. A procedure for obtaining accurate eigenvalues of non-self-adjoint (as well as self-adjoint) equations without introducing adjoint functions is developed. Several numerical examples are provided to illustrate the points made in the chapter, particularly with regard to equations with constraints.

The least-squares and conventional variational approaches are then applied to inhomogeneous equations and the respective results compared. Of particular interest is the case (Fourier series solution) in which the conventional variational solution has a minimum-square-error-in-the-function interpretation.

In Chapter 4, the least-squares approach is applied to the problem of fuel depletion in a nuclear reactor. This problem is described by nonlinear, constrained, non-self-sufficient equations in several dependent and independent variables and is difficult to treat with conventional methods. Two sets of trial functions are used. Each set has the initial flux distribution as one trial function. For the second spatial trial function, one set uses the end-of-life distribution obtained by a standard depletion computer code calculation, while the second set uses a distribution generated by a perturbation technique. Results obtained with each set compare very favorably with the flux, fuel, and poison solutions given by a standard depletion calculation.

Suggestions are given in Chapter 5 for extending the present fuel depletion study to more complicated cases and for further development of generalized variational methods.

The significance of the adjoint function from a general point of view is discussed in Appendix A, where the "importance" concept is extended to nonlinear problems. An understanding of adjoint functions is helpful in the selection of adjoint trial functions and provides additional information about the intrinsic properties of the system considered.

Chapter 2

VARIATIONAL AND OTHER TRIAL-FUNCTION METHODS

This chapter is a review of the principles, objectives, and limitations of variational and other trial-function methods. An acquaintance with the basic principles of the calculus of variations is assumed. A review of these principles is provided in Appendix B. In addition, some references are made to the significance which can be assigned to adjoint functions, as discussed in Appendix A.

Section 2.1 discusses the types of problems which can be treated by variational methods. The basic principles underlying variational methods are discussed in Section 2.2. Section 2.3 treats the problem of determining a functional that yields a desired Euler equation.

In Section 2.4, the manner in which variational methods fit into the framework of the method of weighted residuals, the most general trial-function method, is discussed; the weighted residual approach is extended so as to include constrained equations.

The selection of trial function is discussed in Section 2.5, where it is shown how the interpretation of the adjoint function may be used to generate adjoint trial functions. Also discussed in this section is the question of how to evaluate the merits of the various weighted residual techniques. An important problem to be faced in attempting to resolve this question is seen to be the establishment of a suitable definition of the "best" approximation to the solution.

In Section 2.6, various difficulties that arise in the use of variational methods are discussed. The subsequent chapters in this work are concerned with proposing and analyzing improved methods that are not hindered by these difficulties.

In Section 2.7, it is shown how one of these difficulties, the inability to treat initial-value problems with self-sufficient methods, can be overcome in certain cases.

2.1 The Types of Problems Considered

The following is the basic problem with which we are concerned: Given the equation, which does not have a known exact solution,

$$H\phi - f = 0 \qquad (2.1)$$

3

and a set of N "trial functions" ϕ_i, which are considered to be capable of describing the unknown function ϕ, find the N parameters a_i such that

$$\sum_{i=1}^{N} a_i \phi_i \tag{2.2}$$

is the "best" approximation that can be made for the function ϕ with the trial functions ϕ_i. (In multidimensional problems, the ϕ_i are sometimes taken to be functions of some of the independent variables, and the a_i become functions of the other independent variables.)

All trial-function methods are concerned with this type of problem. Variational methods, however, have additional applications because a variational procedure yields (in addition to this approximate solution) an accurate value for some quantity related to the problem. This fact is important, because we often are more interested in a single characteristic quantity than in the function itself. In quantum theory, for example, we may wish to know the energy level but may not be too concerned about the details of the wave function.

An important general application of the single-accurate-quantity property is in the determination of eigenvalues. Given the equation

$$H\phi - \lambda M\phi = 0 \tag{2.3}$$

it is well known[16] that the functionals

$$\lambda = \frac{\displaystyle\int_a^b dx\ \phi H\phi}{\displaystyle\int_a^b dx\ \phi M\phi} \tag{2.4}$$

in the case where H and M are self-adjoint operators, and

$$\lambda = \frac{\displaystyle\int_a^b dx\ \phi^* H\phi}{\displaystyle\int_a^b dx\ \phi^* M\phi} \tag{2.5}$$

in the non-self-adjoint case, are stationary with respect to variations in the function ϕ and in the adjoint function ϕ^*.

A somewhat related application arises in linear systems in connection with survey calculations. Suppose that we are dealing with
a system described by the equation

$$H\phi - f = 0 \qquad\qquad (2.6)$$

with which is associated a "figure of merit," or characteristic of
performance, of the form

$$J = \int_a^b dx\ g(x)\ \phi(x) \qquad\qquad (2.7)$$

Suppose, further, that we are interested in evaluating this figure
of merit for various "sources" f, but that we are not overly concerned with the detailed behavior of $\phi(x)$. We would then attempt
to express J in terms of f(x) and a Green's function or transfer
function (two common terms used to connote the response to a
unit source) in the form

$$J = \int_a^b dx\ T(x)\ f(x) \qquad\qquad (2.8)$$

It is shown in Appendix A that T(x) is simply the adjoint function
$\phi^*(x)$ given by

$$H^*\phi^* - g = 0 \qquad\qquad (2.9)$$

the equation that arises in the dual-functional approach (see Section 2.3).

The interpretation of the adjoint function as a Green's function
has another implication that encourages the use of dual-functional
variational methods. Because the adjoint function is the contribution of a unit source to the figure of merit, the adjoint function
provides insight into how the source might be modified in order to
improve the figure of merit. The engineer thus learns not only
how his system will behave (from the function) but also how it may
be improved (from the adjoint function). For further details, see
Appendix A.

2.2 Variational Methods

Let us see how the calculus of variations may be used to solve
the basic problem posed in Section 2.1, i.e., to find the parameters a_i.

Suppose we know of a functional

$$J = \int_a^b dx\ F(x, \phi) \tag{2.10}$$

whose Euler equation is the equation of interest, 2.10. Let us substitute the Expansion 2.2 for ϕ in Equation 2.10, thus yielding

$$J = \int_a^b dx\ F(x, a_1, a_2, \cdots, a_N) \tag{2.11}$$

Since J is insensitive to arbitrary variations in ϕ, it must be insensitive to arbitrary variations in the parameters a_i, thus implying the N equations

$$\frac{\partial J}{\partial a_i} = 0 \quad i = 1, 2, \cdots, N \tag{2.12}$$

We therefore obtain a solution to Equation 2.1 and also get a value of J which has second-order accuracy, i.e., if J_0 is the exact value of the functional; and if $\delta\phi(x)$ is the error in the approximate solution for ϕ, then

$$J = J_0 + O\left[\int_a^b dx\ (\delta\phi)^2\right] \tag{2.13}$$

If we are dealing with a multidimensional problem, then we might prefer a "semidirect" approach. In two dimensions, for example, we might replace Equation 2.2 with

$$\sum_{i=1}^N u_i(y)\ \phi_i(x) \tag{2.14}$$

Substituting into a functional

$$J = \int_a^b dx \int_g^h dy\ F(x, y, \phi) \tag{2.15}$$

we have

$$J = \int_a^b dy\ F(y, u_1, u_2, \cdots, u_N) \tag{2.16}$$

Setting the variations of J with respect to the u_i to zero, we have a set of ordinary differential equations for the u_1.

2.3 Construction of Variational Principles

2.3.1 "Self-sufficient" Equations. In order to implement variational methods, it is necessary to obtain the functional that yields Equation 2.1 as its Euler equation. If such a functional exists that does not require the introduction of additional functions (as in Section 2.3.2), then the equation will be referred to as self-sufficient.† If we assume the existence of this functional, its first variation must be

$$\delta J = \int_a^b dx\ \delta\phi(H\phi - f) \tag{2.17}$$

The variational principle can then be formed by transforming Equation 2.17 into

$$\delta J = \delta \int_a^b dx\ F(x, \phi) \tag{2.18}$$

As an example of how this may be done, consider the equation

$$\frac{d}{dx}\left(p\,\frac{d\phi}{dx}\right) + q\phi - f = 0 \tag{2.19}$$

The first variation is

$$\delta J = \int_a^b dx\ \delta\phi\left[\frac{d}{dx}\left(p\,\frac{d\phi}{dx}\right) + q\phi - f\right] \tag{2.20}$$

Note that

---

†In linear problems the term self-adjoint is often used in the literature. Because the adjoint of a nonlinear operator cannot be the same as that operator (as is shown in Appendix A), the term self-adjoint cannot be used for nonlinear equations.

$$\delta\phi\,\frac{d}{dx}\left(p\,\frac{d\phi}{dx}\right) = -\delta\left[\frac{1}{2}p\left(\frac{d\phi}{dx}\right)^2\right] + \frac{d}{dx}\left[p\,\frac{d\phi}{dx}\,\delta\phi\right] \qquad (2.21)$$

$$\delta\phi\,q\phi = \delta\left(\frac{1}{2}q\phi^2\right) \qquad (2.22)$$

$$\delta\phi\,f = \delta(f\phi) \qquad (2.23)$$

We may therefore write the first variation in the form

$$\delta J = \delta\int_a^b dx\left[-\frac{1}{2}p\left(\frac{d\phi}{dx}\right)^2 + \frac{1}{2}q\phi - f\phi\right] + p\,\frac{d\phi}{dx}\,\delta\phi\,\Big|_a^b \qquad (2.24)$$

If the boundary conditions are such that the second term on the right side of Equation 2.24 is zero, then Equation 2.25 is of the form of Equation 2.18 and the desired variational principle has been obtained.

A variational principle is easily found when H in Equation 2.1 is a linear self-adjoint operator, that is, when

$$\int_a^b dx\,(\delta\phi)\,H\phi = \int_a^b dx\,\phi H\delta\phi = \frac{1}{2}\delta\int_a^b dx\,\phi H\phi \qquad (2.25)$$

The operator

$$\frac{d}{dx}\,p\,\frac{d}{dx}$$

in Equation 2.19 is such an operator. A variational principle for Equation 2.1 would then be

$$\delta J = \int_a^b dx\left(\frac{1}{2}\phi H\phi - f\phi\right) = 0 \qquad (2.26)$$

The property of self-adjointness can be useful in some nonlinear problems too. Consider the equation

$$L\phi - f(\phi, x) = 0 \qquad (2.27)$$

where L is linear and self-adjoint, the nonlinearity arising in f. If

$$g(\phi, x) = \int^{\phi} d\phi' \, f(\phi', x) \tag{2.28}$$

then a suitable functional is

$$J = \int_a^b dx \left[\frac{1}{2} \phi L \phi - g(\phi, x) \right] \tag{2.29}$$

2.3.2 Dual Functionals and Adjoint Functions. If we are unable to find a self-sufficient variational principle, we can introduce an auxiliary or adjoint function ϕ^*, the variation of which yields the desired equation. The first variation is

$$\delta J = \int_a^b dx \, \delta \phi^* (H\phi - f) \tag{2.30}$$

Integration with respect to ϕ^* yields the functional

$$J = \int_a^b dx [\phi^* (H\phi - f) - h(\phi, x)] \tag{2.31}$$

where h is the "constant" of integration. With the adjoint operator H^* defined by

$$\int_a^b dx \, \phi^* \delta(H\phi) = \int_a^b dx \, \delta \phi H^* \phi^* \tag{2.32}$$

variation with respect to ϕ yields the equation

$$H^*\phi^* - \frac{\partial h}{\partial \phi} = 0 \tag{2.33}$$

Usually, h is taken to be $g(x)\phi$, and Equation 2.33 becomes the conventional adjoint equation

$$H^*\phi^* - g = 0 \tag{2.34}$$

The relation Equation 2.32 implies adjoint boundary conditions of the form

$$P(\phi, \phi^*, x, \delta\phi, \delta\phi^*)\Big|_a^b = 0 \tag{2.35}$$

In linear problems P is called the bilinear concomitant.[17]

The introduction of the adjoint function requires a second trial-function expansion

$$\phi^*(x) = \sum_{i=1}^{N} a_i^* \phi_i^*(x) \tag{2.36}$$

In addition to Equation 2.14, we obtain the equation

$$\frac{\partial J}{\partial a_i^*} = 0 \quad i = 1, 2, \cdots, N \tag{2.37}$$

In inhomogeneous problems, there is a certain amount of flexibility in the definition of the adjoint equation, since we have a free choice of the adjoint inhomogeneous term g. This flexibility can be used to advantage, as is shown in Appendix A. In linear homogeneous (eigenvalue) problems, there are no inhomogeneities; a given equation implies a specific adjoint equation.

2.4 The Method of Weighted Residuals

2.4.1 The General Approach. Variational methods are not the only procedures that attempt to produce satisfactory solutions from trial functions. Other commonly used methods are Galerkin's method, the method of collocation, and the method of least squares. All these methods, however, are special cases of a general approach called the method of weighted residuals.

Consider the equation

$$H\phi - f = 0 \tag{2.38}$$

and the trial-function expansion

$$\phi = \sum_{i=1}^{N} a_i \phi_i(x) \tag{2.39}$$

The residual is defined by

$$R(x) = \sum_{i=1}^{N} a_i H\phi_i(x) - f(x) \tag{2.40}$$

The N parameters a_i are determined by the N equations

$$\int_a^b dx \, w_j(x) \, R(x) = 0 \qquad j = 1, 2, \cdots, N \qquad (2.41)$$

where the w_j are weighting functions. The various trial function techniques are distinguished by the choice of the w_j.

2.4.2 Variational and Galerkin Methods. Variational methods require either

$$\int_a^b dx \, \delta \phi(H\phi - f) = 0 \qquad (2.42)$$

or

$$\int_a^b dx \, \delta \phi^*(H\phi - f) = 0 \qquad (2.43)$$

depending on whether or not Equation 2.38 is self-sufficient. When the function is expanded as in Equation 2.39 and the adjoint function as in

$$\phi^*(x) = \sum_{i=1}^N a_i^* \phi_i^*(x) \qquad (2.44)$$

the variations, Equations 2.42 and 2.43 become the sets of equations

$$\int_a^b dx \, \phi_j(x) \, R(x) = 0 \qquad j = 1, 2, \cdots, N \qquad (2.45)$$

$$\int_a^b dx \, \phi_j^*(x) \, R(x) = 0 \qquad j = 1, 2, \cdots, N \qquad (2.46)$$

Comparison with Equation 2.41 indicates that variational methods are weighted residual methods with weighting functions ϕ_j or ϕ_j^*.

Galerkin's method is to apply Equation 2.45 regardless of whether or not the equation is self-sufficient.

2.4.3 The Method of Collocation.

This approach requires that the equation be satisfied exactly at N points, i.e., that

$$R(x_j) = 0 \quad j = 1, 2, \cdots, N \tag{2.47}$$

The weighting functions are thus

$$w_j(x) = \delta(x - x_j) \tag{2.48}$$

where $\delta(x - x_j)$ is the Dirac delta function.[18] This method has the advantage that the integrations involved are trivial. However, a serious disadvantage arises because forcing exact agreement at selected points can result in severe fluctuations in the regions between these points.

2.4.4 The Method of Least Squares.

This method aims at minimizing the integrated-square residual

$$I = \int_a^b dx\, p(x)\, R^2(x) \tag{2.49}$$

where $p(x)$ is an arbitrary positive function. The trial parameters a_j are determined by

$$\frac{\partial I}{\partial a_j} = 0 \quad j = 1, 2, \cdots, N \tag{2.50}$$

The weighting functions are

$$w_j(x) = p(x)\, \frac{\partial}{\partial a_j}\, R(x) \tag{2.51}$$

2.4.5 Eigenvalue Problems.

The use of nonvariational methods for eigenvalue problems can lead to difficulty. Consider the equation

$$L\phi - \lambda M\phi = 0 \tag{2.52}$$

When we introduce Expression 2.39, the residual is

$$R(x) = \sum_{i=1}^N a_i(H\phi_i - \lambda M\phi_i) \tag{2.53}$$

Using a set of weighting functions w_j, we are led to a set of homogeneous algebraic equations. Eigenvalues are found by setting the determinant of these equations equal to zero, that is, from

$$|\underset{\sim}{C} - \lambda \underset{\sim}{D}| = 0 \tag{2.54}$$

where the matrix elements are

$$C_{ji} = \int_a^b dx\, w_j L\phi_i \tag{2.54a}$$

$$D_{ji} = \int_a^b dx\, w_j M\phi_i \tag{2.54b}$$

If L and M are self-adjoint operators, then the eigenvalues of Equation 2.52 are real.[4] For Equation 2.54 to guarantee real eigenvalues, however, the matrices $\underset{\sim}{C}$ and $\underset{\sim}{D}$ should be symmetric[10] (for example, $C_{ji} = C_{ij}$). This will be the case only if the weighting functions are proportional to the trial functions, that is, if

$$w_j = h(x)\, \phi_j \tag{2.55}$$

Of these methods (Sections 2.4.1 to 2.4.6), only the variational and Galerkin techniques (which are the same in this case) have this property.

These variational equations can be derived from the stationary principles for eigenvalues cited in Section 2.1. We therefore have additional incentive for using variational methods in eigenvalue problems.

2.4.7 Extension to Constrained Equations. The method of weighted residuals is usually applied to unconstrained equations.[3] The method is extended here to include equations subject to constraints.

Many equations involve control parameters that are chosen so as to satisfy specified constraints. If λ is a control parameter, then

$$H(\lambda)\, \phi - f(x, \lambda) = 0 \tag{2.56}$$

subject to

$$\int_a^b dx\, g(\phi) - K = 0 \tag{2.57}$$

is an example of such a problem. The parameter λ is to be chosen such that Equation 2.57 is satisfied. A trial solution

$$\phi = \sum_{i=1}^{N} a_i \phi_i(x) \qquad (2.58)$$

is selected and the parameters a_i determined.

There are two approaches that may be taken. One of these is to solve for the a_i in terms of λ and then to solve for λ in terms of the a_i in Equation 2.57. In such a procedure, Equation 2.56 is satisfied approximately, but the constraint Equation 2.57 is satisfied exactly. The second approach permits the constraint to be satisfied approximately. The latter procedure is not necessarily less accurate than the former one because, in general, we do not necessarily improve results by forcing strict conformity to a given condition by an already inherently approximate solution. The exactly satisfied constraint will be considered first.

The use of Equation 2.58 in Equation 2.56 yields the residual

$$R(x, \lambda) = \sum_{i=1}^{N} a_i H(\lambda) \, \phi_i - f(x, \lambda) \qquad (2.59)$$

We then apply N weighting functions w_j to obtain the N weighted residual equations

$$\int_0^a dx \, w_j R(x, \lambda) = 0 \qquad j = 1, \cdots, N \qquad (2.60)$$

in N + 1 unknowns. Use of Equation 2.58 in Equation 2.57 yields

$$\int_a^b dx \, g\left(\sum_{i=1}^{N} a_i \phi_i \right) - K = 0 \qquad (2.61)$$

providing the (N + 1)-th equation.

In general, this procedure is straightforward and may be used for any constraint and any control variable that permit a solution. This is not the case, however, for the variational method. The Lagrange multiplier technique (Appendix B) indicates that if the problem is to be self-sufficient, then the λ dependence in Equation 2.56 must be in a term of the form

$$\lambda \frac{\partial g}{\partial \phi} \qquad (2.62)$$

This is a serious limitation of the flexibility of variational methods. It will be discussed in greater detail in Section 2.6.

If the constraint is not to be satisfied exactly, then the constraint equation may be considered as just an additional equation that is to be satisfied in the weighted residual sense. The control parameter λ is then considered to be a variable in the same sense as is ϕ. The method of Section 2.4.1 is then applied assuming that Equation 2.38 is a matrix equation (a set of simultaneous equations), with H a matrix operator and ϕ and f unknown vectors. The ϕ_i and w_i become trial vectors and weighting vectors, and Equation 2.41 is a set of scalar products.

2.5 The Selection of Trial and Weighting Functions

2.5.1 Use of Physical and Mathematical Intuition for Trial Functions.
The ability to select adequate trial functions is necessary for the successful utilization of variational and other weighted residual methods. Quite often, simple physical and mathematical reasoning yields a great deal of information about the form of the solution, thus aiding in the task of generating trial functions.

Consider the illustrative example of a body of mass m undergoing gravitational acceleration from an initial velocity v_0 in an atmosphere with a quadratic drag force. The velocity v is given by

$$m \frac{dv}{dt} = mg - Kv^2 \qquad v(0) = v_0 \qquad (2.63)$$

where g is the (constant) gravitational acceleration and K is a constant. Physically, it is clear that the body will approach a terminal velocity v_T, at which the forces of drag and gravity are balanced, given by

$$v_T = \sqrt{\frac{mg}{K}} \qquad (2.64)$$

To estimate the order of magnitude of the time required for the body to attain essentially this velocity, we may analyze dimensionally the parameters m, g, and K to yield the time parameter

$$\tau = \sqrt{\frac{m}{Kg}} \qquad (2.65)$$

Therefore, we choose a trial solution that not only satisfies the initial condition but also approaches the terminal velocity v_T with relaxation time τ. A suitable trial function might be

$$v_0 + (v_T - v_0) \tanh\left(\frac{t}{\tau}\right) \qquad (2.66)$$

As a second example, let us suppose that the equation governing a certain process can be solved only in "extreme" cases. We might be able to show mathematically, or we might feel intuitively, that the solution in intermediate cases can be expressed in terms of the extreme solutions. We might then use these extreme solutions as trial functions and use a weighted residual technique to "interpolate."

2.5.2 Adjoint Trial Functions. In dual problems, we are required to generate adjoint trial functions. This can be a major source of difficulty, since we are usually not as familiar with the adjoint equation as with the original equation. Reynolds[24] cited this difficulty in his study of the application of variational methods to laminar-flow problems. In Appendix A a study is made of the significance of the adjoint function. Let us consider how this significance may be used to generate information about the solution to the adjoint problem. The procedure is illustrated by reference to the neutron slowing-down problem.

The neutron flux as a function of energy is given by the integral equation

$$\Sigma_t(E) \, \phi(E) \; = \; \int_0^\infty \Sigma_s(E' \to E) \, \phi(E') \, dE' + S(E)$$

$$(2.67)$$

where $\Sigma_t(E)$ is the total reaction cross section and $\Sigma_s(E' \to E)$ is the cross section for scattering from energy E' to energy E. With the adjoint source chosen as the absorption cross section, the adjoint equation becomes (according to the development by Morse and Feshbach[20] on obtaining adjoints of integral equations)

$$\Sigma_t(E) \, \phi^*(E) \; = \; \int_0^\infty \Sigma_s(E \to E') \, \phi^*(E') \, dE' + \Sigma_a(E)$$

$$(2.68)$$

Following the procedures of Appendix A, the value of the functional may be written either as

$$-J \; = \; \int_0^\infty \Sigma_a(E) \, \phi(E) \, dE \qquad\qquad (2.69)$$

or as

$$-J \; = \; \int_0^\infty S(E) \, \phi^*(E) \, dE \qquad\qquad (2.70)$$

and the adjoint flux $\phi^*(E)$ is the total amount of absorption caused by a unit source at E. Since this is a steady-state problem, every source neutron must be absorbed, so

$$\phi^*(E) = 1 \tag{2.71}$$

This solution can be verified easily. Substitution into the adjoint equation yields

$$\Sigma_t(E) = \Sigma_s(E) + \Sigma_a(E) \tag{2.72}$$

where $\Sigma_s(E)$, the total scattering cross section, is given by

$$\Sigma_s(E) = \int_0^\infty dE' \, \Sigma_s(E \to E') \tag{2.73}$$

The cross-section relation Equation 2.72 is satisfied identically.
    In this case, the entire energy range was considered. Quite often, however, the slowing-down problem is subdivided into two problems, epithermal slowing down and thermalization. In the epithermal region, scattering is assumed to take place only from higher to lower energies. The slowing-down equation in the epithermal region becomes

$$\Sigma_t(E) \, \phi(E) = \int_E^\infty \Sigma_s(E' \to E) \, \phi(E') \, dE' + S(E) \tag{2.74}$$

and the adjoint equation becomes (following Reference 20 for Volterra integral equations)

$$\Sigma_t(E) \, \phi^*(E) = \int_{E_T}^E \Sigma_s(E \to E') \, \phi^*(E') \, dE' + \Sigma_a(E) \tag{2.75}$$

where E_T is the "boundary" between the epithermal and thermal regions. The value of the functional is

$$-J = \int_{E_T}^\infty dE \, \Sigma_a(E) \, \phi(E) \tag{2.76}$$

and $\phi^*(E)$ is the amount of absorption in the epithermal region caused by a unit source at E. Since some of the neutrons slow down into the thermal region,

$$\phi^*(E) < 1 \tag{2.77}$$

It is well known[9] that an approximate solution of the slowing down equation is

$$\phi(E) = \frac{C_1}{E} \tag{2.78}$$

where C_1 is a constant. Absorption is governed approximately by the "1/v law,"

$$\Sigma_a(E) = \frac{C_2}{\sqrt{E}} \tag{2.79}$$

where C_2 is a constant. If we use $C = C_1 C_2$, the absorption rate is of the form

$$\Sigma_a(E) \, \phi(E) = C E^{-3/2} \tag{2.80}$$

Integrating from E_T to the energy of the source, we have

$$\int_{E_T}^{E} dE' \, CE'^{-3/2} = \frac{C}{2}\left(\frac{1}{\sqrt{E_T}} - \frac{1}{\sqrt{E}}\right) \tag{2.81}$$

Therefore, a trial function for the adjoint flux might be

$$\frac{1}{\sqrt{E_T}} - \frac{1}{\sqrt{E}} \tag{2.82}$$

It can be seen by inspection of Equation 2.75 that the adjoint flux satisfies the boundary condition

$$\phi^*(E_T) = \frac{\Sigma_a}{\Sigma_t}(E_T) \tag{2.83}$$

Combining Equations 2.82 and 2.83, we obtain the trial solution

$$\phi^*(E) = \frac{\Sigma_a}{\Sigma_t}(E_T) + A\left(\frac{1}{\sqrt{E_T}} - \frac{1}{\sqrt{E}}\right) \tag{2.84}$$

and A is the parameter to be determined.

It sometimes happens that the adjoint equation is of such a form that the trial functions used for the original equation can also be used for the adjoint equation. Consider for example, the initial-value problem

$$\frac{\partial \phi}{\partial t} = \frac{\partial^2 \phi}{\partial x^2} + \lambda \phi \qquad \phi(0, x) = \phi_0(x) \tag{2.85}$$

The adjoint equation is the final-value problem

$$-\frac{\partial \phi^*}{\partial t} = \frac{\partial^2 \phi^*}{\partial x^2} + \lambda \phi^* \qquad \phi^*(T, x) = \phi_T^*(x) \tag{2.86}$$

Suppose the adjoint final value condition is chosen to be

$$\phi_T^*(x) = \phi_0(x) \tag{2.87}$$

Putting $\tau = T - t$ in Equation 2.86, we have

$$\frac{\partial \phi^*}{\partial \tau} = \frac{\partial^2 \phi^*}{\partial x^2} + \lambda \phi^* \qquad \phi^*(0, x) = \phi_0(x) \tag{2.88}$$

that is, the adjoint equation becomes the same as Equation 2.85. A trial-function expansion

$$\phi(x) = \sum_{i=1}^{N} a_i \phi_i(x) \tag{2.89}$$

which is valid for Equation 2.85 should therefore be valid also for Equation 2.86.

2.5.3 Error Criteria and Weighting Functions. All trial-function methods consist of making a residual orthogonal to a set of weighting functions. A logical question to ask is, "What choice of weighting functions yields the "best" solution?" Before attempting to answer this question, we must first define what we mean by "best." Such a definition is often difficult to make because we have usually only a vague requirement that the approximate solution should be a "good" representation of the exact solution.

A precise definition might be that the integrated, squared deviation from the exact solution

$$\int_a^b dx \left(\phi - \sum_{i=1}^{N} a_i \phi_i \right)^2 \tag{2.90}$$

be as small as possible. However, since the exact solution is not known, this definition is not readily applicable.

Roussopolos-type variational methods,[25] though, using the functional

$$J = \int_a^b dx \, (\phi^* H\phi - f\phi^* - g\phi) \tag{2.91}$$

yield approximate solutions such that an accurate value of a linear-weighted integral of the function

$$J = -\int_a^b dx \, g(x) \, \phi(x) \tag{2.92}$$

is obtained. In this sense, variational methods tend to reduce the error in the function. Because the integrand of Equation 2.91 is not positive definite, however, the accuracy in the quantity J can be achieved to some extent by cancellation of errors, a matter discussed in more detail in Section 2.6.

Another possible standard of comparison is the residual itself. The residual, as defined in Section 2.4, can be considered the error in the equation. Procedures that are particularly effective in making the error in the equation small might be considered "superior" methods. The method of least squares (Section 2.4.4) is clearly such a method.

Variational methods have this property when dealing with linear homogeneous equations, using functionals

$$I_1 = \int_a^b dx \, \phi L\phi \tag{2.93}$$

$$I_2 = \int_a^b dx \, \phi^* L\phi \tag{2.94}$$

for self-adjoint and non-self-adjoint operators, respectively. Variational methods thus are seen to yield accurate function-weighted and adjoint-weighted residuals in linear homogeneous problems.

2.6 Limitations of Conventional Variational Methods

2.6.1 Self-sufficiency and Versatility. In many cases, even though the equation of interest can be derived from a self-sufficient

variational principle, other factors arise that interfere with the use of such a principle. Such factors enter when boundary conditions are applied, when constraints are specified, and when attempts are made at rewriting the equation in a somewhat different form because of computational convenience. These matters will be discussed in the order mentioned.

To illustrate the problem caused by boundary conditions, consider the equation

$$\frac{d^2\phi}{dx^2} + k^2\phi = f(x) \tag{2.95}$$

The functional

$$I = \int_a^b dx \left[\left(\frac{d\phi}{dx}\right)^2 - k^2\phi + 2f\phi\right] \tag{2.96}$$

yields Equation 2.95, provided the boundary term

$$\left(\frac{d\phi}{dx}\delta\phi\right)\Big|_a^b = 0 \tag{2.97}$$

vanishes. If the problem under consideration is a boundary-value problem, for example, if Equation 2.95 is the neutron-diffusion equation with the flux specified at the boundaries, then Equation 2.97 can be satisfied. If, on the other hand, the problem is an initial-value problem, for example, the behavior of an harmonic oscillator, then only the lower limit of Equation 2.97 can be made to vanish. Since Equation 2.97 cannot be satisfied in an initial-value problem, Equation 2.96 cannot be said to yield Equation 2.95. This is an important point, because many engineering and scientific problems are initial-value problems.

To illustrate the problem caused by constraints, consider again Equation 2.95 with k^2 a parameter to be chosen such that

$$\int_a^b dx\, g(\phi) = K \tag{2.98}$$

The standard variational approach is to apply the method of Lagrange multipliers (as discussed in Appendix B). The "control parameter" k^2 is the Lagrange multiplier for the problem. We set up the functional

$$I = \int_a^b dx \left[\frac{1}{2} \left(\frac{d\phi}{dx} \right)^2 + f\phi - k^2 g(\phi) \right] \tag{2.99}$$

which yields

$$\frac{d^2\phi}{dx^2} + k^2 \frac{\partial g(\phi)}{\partial \phi} = f(x) \tag{2.100}$$

The Lagrange multiplier approach yields the desired equation only when

$$\frac{\partial g(\phi)}{\partial \phi} = \phi \tag{2.101}$$

that is, only when

$$g(\phi) = \frac{1}{2} \phi^2 \tag{2.102}$$

The approach is consequently applicable for only a specific type of constraint, Equation 2.102. Quite often, however, a different type is of greater interest. In nuclear engineering, for example, a common constraint is that the system operate at a given power level. This would correspond to

$$g(\phi) = \phi \qquad \int_a^b dx \, \phi = K \tag{2.103}$$

To illustrate the problem of convenience of form, consider the equation

$$\frac{d^2\phi}{dx^2} + a\phi + b \frac{\phi^2}{\phi + c} = 0 \tag{2.104}$$

where a, b, and c are constants. The neutron-diffusion equation with xenon poisoning considered is of this form. As was shown in Section 2.3, nonlinear equations in which the operators are linear and self-adjoint can be derived from self-sufficient functionals. With Equation 2.104, however, a logarithmic term of the form

$$\ln (\phi + c) \tag{2.105}$$

will appear in the integrand of the functional. This introduces a serious difficulty. It may not be possible to integrate Equa-

tion 2.105 with respect to x analytically. On the other hand, the
function ϕ, having been expanded in a series of trial solutions,
contains unknown parameters. This fact stands in the way of
evaluating the integral numerically.

Logarithmic terms might be avoided by rewriting Equation 2.105 in the form

$$(\phi + c) \frac{d^2\phi}{dx^2} + a(\phi + c)\phi + b\phi^2 = 0 \qquad (2.106)$$

In this form, however, the first term is not self-adjoint, and
Equation 2.106 cannot be obtained from a self-sufficient principle. Thus, merely rewriting the equation in a slightly different form precludes its derivation from a self-sufficient variational principle.

Therefore, in several respects, self-sufficient variational
methods lack versatility. Whether or not these methods are applicable to a given equation depends on the nature of the boundary
conditions, on the type of constraint, and even on the form in which
the equation is written. When a self-sufficient method cannot be
used, variational methods may still be applied if dual functionals
are invoked. Such an alternative introduces problems of its own,
as will be shown in the following section.

2.6.2 Problems with Dual Functionals. When dual functionals
are used, we must concern ourselves with the problem of generating adjoint trial functions. In selecting trial functions, the
analyst may utilize the properties of adjoint functions, as illustrated in Section 2.5. It is possible, however, that despite the
interpretation that can be given to adjoint functions, it may be difficult to construct adjoint trial functions which are of the same
caliber as are the "system" trial functions. One reason for this
is because there is, generally, less familiarity with adjoint functions than with "system" functions.

As an illustration, suppose that an engineer has been assigned
the task of determining the flux distribution in space and energy
for a certain type of nuclear reactor. As an aid in generating
trial functions, he may take advantage of the state of the art,
for example, he may consult the literature describing flux distributions in similar reactors. Not nearly as much information
is available concerning adjoint functions. It is, therefore, often
much easier to generate system trials than to obtain adjoint
trials.

Another source of difficulty is that in constrained problems
the adjoint equation is more complicated than the system equation. This is because the dual-functional approach requires the
specification of an additional constraint that involves the adjoint
functions, as is easily shown. The constraint

$$\int_a^b dx \, G(\phi) \, dx = 0 \tag{2.107}$$

causes the dual functional to be modified to

$$J = \int_a^b dx \left[\phi^*(H\phi - f) - g\phi - \lambda G(\phi) \right] \tag{2.108}$$

where λ is the Lagrange multiplier. Varying ϕ and ϕ^* yields

$$H\phi - f = 0 \tag{2.109}$$

$$H^*\phi^* - g - \lambda \frac{\partial G}{\partial \phi} = 0 \tag{2.110}$$

A constraint involving only the system function ϕ affects only the adjoint Equation 2.110. In order to introduce a Lagrange multiplier into the system Equation 2.109, we must provide a constraint affecting the adjoint function.

As an illustration of how the adjoint equation is affected, consider the equation

$$H\phi + k_1\phi^2 = 0 \tag{2.111}$$

where k_1 is a parameter to be determined such that

$$\int_a^b dx \, \phi = c_1 \tag{2.112}$$

This requires an adjoint constraint of the form

$$\int_a^b dx \, \phi^*\phi^2 = c_2 \tag{2.113}$$

The functional to be varied is

$$I = \int_a^b dx \left[\phi^*(H\phi + k_1\phi^2) + k_2\phi \right] \tag{2.114}$$

The multiplier k_2 is introduced through the constraint Equation 2.112 whereas the multiplier k_1 is introduced through the adjoint constraint Equation 2.113. The adjoint equation is

$$H^*\phi^* + 2k_1\phi\phi^* + k_2 = 0 \qquad (2.115)$$

The adjoint equation is seen to involve two Lagrange multipliers. (Equation 2.111 has only one.) If the control variable were a function of x, rather than simply a parameter, the resulting adjoint equation would be still more complicated. Also, matters would be worse in the case of multiple constraints, since the adjoint equation contains two Lagrange multipliers for every one in the system equation.

In nonlinear problems in general, it is difficult to generate adjoint trial functions. This is because the significance of the adjoint function is more subtle in nonlinear problems than in linear problems. In multidimensional problems, as is shown in Appendix A, even this interpretation of the adjoint function may be of limited value.

In some cases the question of whether or not we are able to construct adjoint trial functions is irrelevant. The analyst may simply feel that knowledge of the adjoint function is not sufficiently important to merit the work involved in generating adjoint trial functions. As an illustration, let us consider the synthesis problem.[12] The time consumed in the numerical solution of a boundary-value problem varies exponentially with the number of dimensions involved. Synthesizing two-dimensional solutions by combining one-dimensional solutions can yield approximate solutions in a relatively short time. Consider, for example, the equation

$$(L_x + L_y)\,\phi(x, y) + [\lambda + u(y_0)\,f(x)]\,\phi(x, y) = 0 \quad (2.116)$$

where $u(y_0)$, the unit step function, is one for $y > y_0$ and zero for $y < y_0$. For $y \gg y_0$ we might expect the x dependence to be given by the solution to

$$L_x\phi_1(x) + [\lambda + f(x)]\,\phi_1(x) = 0 \qquad (2.117)$$

and for $y \ll y_0$ by

$$L_x\phi_2(x) + \lambda\phi_2(x) = 0 \qquad (2.118)$$

assuming separability far from the interface y_0. A trial solution might then be

$$\phi(x, y) = u_1(y)\,\phi_1(x) + u_2(y)\,\phi_2(x) \qquad (2.119)$$

The analyst solves Equations 2.117 and 2.118 to obtain the trial functions for Equation 2.120. He then uses a semidirect weighted residual technique to solve for $u_1(y)$ and $u_2(y)$. If this technique is the dual-functional variational method, then the solutions to the adjoint equations

$$L_x^* \phi_1^*(x) + [\lambda + f(x)] \, \phi_1^*(x) = 0 \qquad (2.120)$$

$$L_x^* \phi_1^*(x) + \lambda \phi_1^*(x) = 0 \qquad (2.121)$$

are needed. There is consequently about twice as much effort required in the use of dual functional approach as in other weighted residual techniques. Kaplan[13] prefers to use the Galerkin method for this reason.

2.6.3 The Problem of Cancellation of Errors. In conventional variational methods, the adjustable parameters were selected in such a way as to yield an accurate value of a single parameter, the functional. It is then assumed that the approximate solution for the function thus obtained is also an accurate one. This assumption, however, is not necessarily valid. It may happen that positive and negative errors may cancel each other in the evaluation of the functional, thus leading to an accurate functional despite an inaccurate function. This possibility tends to reduce the reliability of the approximate solution.

It was pointed out in Section 2.5 that error cancellation can take place in the functional

$$J = \int_a^b dx \, [\phi^*(H\phi - f) - g\phi] \qquad (2.122)$$

Positive and negative errors can occur in each of the terms in the integrand. The same situation prevails when H is self-adjoint and the principle is

$$J = \int_a^b dx \, (\phi H\phi - 2f\phi) \qquad (2.123)$$

Self-adjoint homogeneous problems often are associated with positive definite functional expressions for the eigenvalues, such as

$$\lambda = \frac{\int_a^b dx \left[\left(\frac{d\phi}{dx}\right)^2 + p(x) \, \phi^2 \right]}{\int_a^b dx \, \phi^2} \qquad (2.124)$$

which has the Euler equation

$$\frac{d^2\phi}{dx^2} - p(x) \, \phi + \lambda\phi = 0 \qquad (2.125)$$

In Equations 2.125 and 2.124, $p(x)$ is a positive definite function. Because every term in Equation 2.124 is positive definite, an accurate value of λ cannot be obtained by cancellation of errors.

In general, however, functional expression for eigenvalues are not positive definite. In fact, functionals (such as Equation 2.5) that involve adjoint functions never have this property.[22]

Except in cases of certain self-adjoint eigenvalue problems, the possibility of cancellation of errors exists, thereby serving to impair confidence in the ability of the procedure to yield a satisfactory approximate solution.

2.7 Initial-Value Problems and the Simulated Boundary-Value Problem

In this section we present a technique that can overcome one of the limitations cited in Section 2.6.1, namely that initial-value problems cannot be treated with self-sufficient methods. The technique is applicable to problems where we have some information about solution behavior in the "future," for example, that it is monotonic or periodic.

In many initial-value problems, such information may be inferred on physical or mathematical grounds. Consider, as an example, the problem of reactor kinetics with negative-temperature feedback and with heat removal neglected. The equations are of the form[11]

$$\frac{dn}{dt} = c_1(1 - c_2 T)n \qquad (2.126)$$

$$\frac{dT}{dt} = c_3 n \qquad (2.127)$$

where n and T are neutron density and temperature, respectively, and where c_1, c_2, and c_3 are constants. An increase in reactivity causes an increase in the neutron level (or power level), which causes an increase in temperature, which causes a decrease in reactivity. The neutron level will initially rise, then reach a maximum value, and finally decrease. Therefore, at some time τ in the future, the following must hold

$$\frac{dn}{dt}(\tau) = 0 \tag{2.128}$$

Equations 2.126 and 2.127 may be combined into a single second-order equation for the neutron density of the form

$$n\frac{d^2n}{dt^2} - \left(\frac{dn}{dt}\right)^2 + cn^3 = 0 \tag{2.129}$$

where c is a constant. This equation can be derived from the self-sufficient functional

$$I = \int_0^T dt \left[\left(\frac{1}{n}\frac{dn}{dt}\right)^2 - 2cn\right] \tag{2.130}$$

provided boundary conditions

$$\frac{2}{n^2}\frac{dn}{dt} \delta n \bigg|_0^T = 0 \tag{2.131}$$

can be satisfied. If we specify the initial condition

$$n(0) = n_0 \tag{2.132}$$

we see that δn vanishes at the lower limit. The other condition to be specified for Equation 2.129 is

$$\frac{dn}{dt}(0) = c_1(1 - c_2T_0)n_0 \tag{2.133}$$

where T_0 is the initial temperature. This condition, however cannot satisfy the upper limit of Equation 2.131.

An alternative is to specify the natural boundary condition Equation 2.128. In this case, τ must be left unspecified. It (τ) may be obtained, however, by invoking Equation 2.133 later in the analysis.

The procedure for the simulated boundary-value problem is as follows. We assume a trial solution for $n(t, \tau)$

$$n(t, \tau) = \sum_{i=1}^{N} a_i y_i(t, \tau) \qquad (2.134)$$

such that Equation 2.134 satisfies Equations 2.132 and 2.128. This trial solution is substituted into the functional Equation 2.130, which is then varied with respect to the a_i. The resulting equations determine the a_i in terms of τ. One then invokes Equation 2.133 to solve for τ and a_i.

For reasons of algebraic and computational convenience, this method is tested on the equation

$$n \frac{d^2 n}{dt^2} + \left(\frac{dn}{dt} \right)^2 - kn = 0 \qquad (2.135)$$

(with k a constant) instead of on Equation 2.129. A functional which yields Equation 2.135 is

$$J = \int_0^T dt \left[\left(n \frac{dn}{dt} \right)^2 + \frac{2}{3} kn^3 \right] \qquad (2.136)$$

The initial conditions applied are

$$n(0) = 0 \qquad (2.137)$$

$$\frac{dn}{dt}(0) = 0 \qquad (2.138)$$

Since it can be shown that the solution increases monotonically, the upper boundary condition can be chosen as

$$n(\tau) = 1 \qquad (2.139)$$

The exact solution to this problem with k = 6 is

$$n(t) = t^2 \qquad \tau = 1 \qquad (2.140)$$

Two trial solutions are considered. The first is

$$n(t) \approx \frac{t}{\tau} + a_0 t(t - \tau) \qquad (2.141)$$

The method yields $a_0 = 1$ and $\tau = 1$, thus showing that the correct values of the parameters are obtained if the proper form of the solution is assumed. The second trial solution is

$$n(t) \approx \frac{t}{\tau} + t(t - \tau)(a_0 + a_1 t) \qquad (2.142)$$

The method yields $a_0 = 1$, $\tau = 1$, and $a_1 = 0$; thus showing that when an extraneous form (t^3 in this case) is assumed in addition to the proper form, the extraneous form is rejected.

In addition to being confined to problems about which we have "future knowledge," the simulated boundary-value problem approach has two drawbacks. One drawback is its limited applicability. Only second-order equations can be treated because it is unlikely that we will know more than one condition at a single future time. The other is that when several trial functions are used the variational equations in terms of τ tend to become fairly complicated.

Chapter 3

LEAST-SQUARES VARIATIONAL METHODS

Several limitations of conventional variational methods were pointed out in Chapter 2. This chapter attempts to develop variational methods that do not possess these limitations and can, therefore, be applied to general problems.

Section 3.1 advances a generalized view of variational techniques. We seek a functional whose Euler equation has the solution of the equation of interest as a particular solution. Since such a functional can always be found, every equation is, in effect, self-sufficient, and it is never necessary (although in some cases it may be desirable) to introduce adjoint functions.

Since many self-sufficient equations, having the solution of the equation of interest as a particular solution, exist, we are faced with the problem of choosing the best functional. A set of criteria is put forth that should be satisfied by variational methods. It is shown in Appendix C that the best functional for a general problem should be of the least-squares type.

The next five sections analyze the properties of least-squares variational methods, and the points made are illustrated with numerical examples. The flexibility of the approach in dealing with constrained equations is emphasized. It is shown how we may take advantage of "extraneous" boundary conditions (implied by the higher-order Euler equation) in order to increase the accuracy of the results with little increase in effort. It is shown how the method may be applied to eigenvalue problems, yielding accurate eigenvalues for both self-adjoint and non-self-adjoint equations. The problem of choosing the relative weighting of terms in the functional is analyzed.

Section 3.7 is a comparison between least-squares variational methods and conventional variational methods. The use of a Fourier expansion as the trial solution is especially interesting because of the minimum-square-error-in-the-function interpretation that can be given to the conventional variational results.

3.1 A Generalized View of Variational Methods

3.1.1 Euler Equations of Higher Order. Consider the equation

$$H\phi - f = 0 \tag{3.1}$$

which cannot be derived from a self-sufficient variational principle. Let us operate on this equation with an operator G such that the resulting equation

$$G(H\phi - f) = 0 \qquad\qquad (3.2)$$

can be so derived.

To prove that this is always possible, it is sufficient to show that a specific choice for G always yields a self-sufficient equation. In Appendix C, this will be shown to be the case when G is the adjoint operator H^* defined in Equation 2.32.

Since Equation 3.2 is an equation of higher order, additional boundary conditions must be specified. These extra conditions must be consistent with Equation 3.1 for the variational solution to give the solution of Equation 3.1 and not a solution of Equation 3.2 which does not satisfy Equation 3.1. One manner of choosing consistent conditions is to specify that

$$\frac{d}{dx}(H\phi - f) = 0 \qquad\qquad (3.3)$$

$$\frac{d^2}{dx^2}(H\phi - f) = 0 \qquad\qquad (3.4)$$

and so forth, be satisfied at the boundaries.

It may not be convenient, however, to force trial functions to satisfy these extra conditions, even though the conditions are consistent with the original equation. We would like to be able to apply only those conditions that are associated with the original problem, Equation 3.1. This would be the case if these extra conditions were introduced as natural boundary conditions.

3.1.2 Criteria Desirable in a Variational Method. Since there are, in general, many operators G that can be used in Equation 3.2, it is worthwhile to set up a list of criteria, which the resulting variational principle should satisfy, and then to seek the principle that best fulfills these requirements. These criteria are defined, basically, in order to avoid the shortcomings of conventional variational methods discussed in Chapter 2. The list of criteria is as follows:

1. The procedure should tend to minimize errors in some sense. Since, as was pointed out in Section 2.5, the residual is a suitable measure of the accuracy of the results, the functional should be related to the residual.

2. In order to prevent the significance of second-order accuracy in the residual to be mitigated by cancellation of errors, the integrand of the functional should be definite (that is, should admit values of only one sign).

3. The procedure should be capable of treating initial-value problems.

4. The procedure should be applicable regardless of the form in which the equation is placed, so that this form can be chosen for computational convenience.

5. The procedure should be able to treat problems with rather arbitrary constraints and should yield accurate control variables.

6. The procedure should yield accurate eigenvalues, even in non-self-adjoint equations.

7. The operator G, the functional, and the computational procedure should be as simple as possible.

8. There should be a simple, universally applicable method for obtaining the operator G and the appropriate functional.

9. The procedure should imply natural boundary conditions of the type discussed in Section 3.1.1.

The general approach has already removed the necessity of introducing an adjoint function.

3.2 Least-Squares Variational Principles

It is shown in Appendix C that the desire for the simplest possible functional which yields an accurate positive-definite residual leads to functionals of the least-squares type

$$I = \int_a^b dx\, p(x)(H\phi - f)^2 \tag{3.5}$$

where $p(x)$ is an arbitrary positive function. The Euler equation is

$$H^*[p(x)(H\phi - f)] = 0 \tag{3.6}$$

where H^* is the adjoint operator.

When dealing with more than one equation, it is natural to obtain a functional using the sums of the squares of the several equations; that is, if we have two equations of the form

$$H_{11}\phi_1 + H_{12}\phi_2 = f_1 \tag{3.7}$$

$$H_{21}\phi_1 + H_{22}\phi_2 = f_2 \tag{3.8}$$

we might use the functional

$$I = \int_a^b dx\, [(H_{11}\phi_1 + H_{12}\phi_2 - f_1)^2 + (H_{21}\phi_1 + H_{22}\phi_2 - f_2)^2] \tag{3.9}$$

Equation 3.9 is obtained from Equation 3.5 with $p(x) = 1$ if Equation 3.1 is considered to be a matrix equation.

A slight modification can be introduced if one of these equations is a constraint, for example, if we deal with the system

$$H\phi = f \qquad (3.10)$$

$$\int_a^b dx\, g(\phi) = K \qquad (3.11)$$

We may form a positive-definite functional by adding the square of the constraint residual to the integral of the square of the equation; that is,

$$J = \int_a^b dx\, (H\phi - f)^2 + c^2 \left[\int_a^b dx\, g(\phi) - K \right]^2 \qquad (3.12)$$

where c^2 is an arbitrary positive constant. The problem of choosing an appropriate c^2 will be discussed in Section 3.7. It should be obvious that c^2 is needed at least to make the terms in Equation 3.12 dimensionally consistent, because the square of an integral introduces one more x dimension than does the integral of a square.

In order to be able to satisfy Equation 3.11, H and/or f must contain a control parameter λ. In taking the first variation of Equation 3.12, the parameter λ is to be treated as a function in the same sense as is ϕ, variations being taken with respect to ϕ and λ; that is,

$$\frac{1}{2}\delta J = \delta\lambda \int_a^b dx\, (H\phi - f) \frac{\partial}{\partial\lambda}(H\phi - f)$$

$$+ c^2 \int_a^b dx\, \delta g \left[\int_a^b dx\, g(\phi) - K \right]$$

$$+ \int_a^b dx\, (H\phi - f)[H\delta\phi + (\delta_\phi H)\phi] \qquad (3.13)$$

where δ_ϕ denotes partial variation with respect to ϕ. Since the

variations with respect to each of the variables must vanish in-
dependently, two equations result, namely

$$\int_a^b dx \, (H\phi - f) \frac{\partial}{\partial \lambda} (H\phi - f) = 0 \tag{3.14}$$

$$c^2 \int_a^b dx \, \delta g \left[\int_a^b dx \, g(\phi) - K \right] + \int_a^b dx \, (H\phi - f)[H\delta\phi + (\delta_\phi H)\phi] = 0 \tag{3.15}$$

If

$$\delta g = \frac{\partial g(\phi)}{\partial \phi} \delta\phi \tag{3.16}$$

and if we use the definition (Equation 2.32) of the adjoint operator, Equation 3.15 becomes

$$\int_a^b dx \, \delta\phi \left\{ c^2 \left[\int_a^b dx \, g(\phi) - K \right] \frac{\partial g}{\partial \phi} + H^*(H\phi - f) \right\} = 0 \tag{3.17}$$

so the Euler Equation 3.12 is Equation 3.14 and

$$c^2 \frac{\partial g}{\partial \phi} \left(\int_a^b dx \, g - K \right) + H^*(H\phi - f) = 0 \tag{3.18}$$

The solution of the system of Equation 3.10 and 3.11 is naturally
a particular solution of the system of Equations 3.14 and 3.18.

If, instead of the distributed constraint Equation 3.11, we were
faced with a local constraint of the type

$$M\phi = g(x) \tag{3.19}$$

where M is some operator, then we might use a functional such
as

$$J' = \int_a^b dx \, (H\phi - f)^2 + c_1^2 \int_a^b dx \, (M\phi - g)^2 \tag{3.20}$$

Instead of a single control parameter, a control function $\lambda(x)$ would be required. We would assume a trial solution for the control function

$$\lambda(x) = \sum_{i=1}^{N} b_i \lambda_i(x) \qquad (3.21)$$

and the variation with respect to $\lambda(x)$ would reduce to differentiation with respect to the b_i. The Euler equations are

$$(H\phi - f) \frac{\partial}{\partial \lambda} (H\phi - f) = 0 \qquad (3.22)$$

and

$$c_i^2 M^*(M\phi - g) + H^*(H\phi - f) = 0 \qquad (3.23)$$

There is an essential difference between the Lagrange multiplier technique and the approach to constraints taken here. The Lagrange multiplier technique considers the constraint as a subsidiary condition applied to the equation. The approach taken here considers the constraint as just another equation. The functional is therefore designed to yield both the equation and the constraint in the Euler equations.

It should be noted that the conventional method of least squares mentioned in Chapter 2 fits into the framework of the generalized variational methods discussed in this section. The variational approach to least-squares methods, as proposed in this section, will be seen to lead to more general applications than those for which the conventional approach has been used.

3.3 Some Variational Aspects of Least Squares

In this section we shall consider the first and second variations of the weighted-square residual functional in order to investigate what this functional implies with regard to boundary conditions and extraneous solutions. The first variation of Equation 3.5 is

$$\delta I = \int_a^b dx\, p(x)\, (H\phi - f)\, [H\delta\phi + (\delta H)\phi] \qquad (3.24)$$

which can be converted to the form

$$\delta I = \int_a^b dx \; (\delta \phi) H^*[p(x) \, (H\phi - f)] \qquad (3.25)$$

Going from Equation 3.24 to Equation 3.25, which usually involves integration by parts, introduces boundary terms of the form

$$(\delta \phi) N_1[p(x) \, (H\phi - f)] \Big|_a^b = 0 \qquad (3.26)$$

$$\left[\frac{d}{dx} (\delta \phi) \right] N_2[p(x) \, (H\phi - f)] \Big|_a^b = 0 \qquad (3.27)$$

and so forth, the operators N_1 and N_2 depending on the form of the operator H.

It is clear from Equations 3.26 and 3.27 that the "extra" natural boundary conditions introduced by the higher-order Equation 3.25 are consistent with the original Equation 3.1, and are of the general type called for in Section 3.4. Thus, another of the requirements of Section 3.1 is satisfied.

Taking the variation of Equation 3.24 yields the second variation

$$\delta^2 I = \int_a^b dx \; p(x) \, [H\delta \phi + (\delta H)\phi]^2$$

$$+ \int_a^b dx \; p(x) \, (H\phi - f) \, [2(\delta H)(\delta \phi) + (\delta^2 H)\phi]$$

$$\qquad (3.28)$$

Note that if H is a linear operator, then Equation 3.28 becomes

$$\delta^2 I = \int_a^b dx \; p(x) \, [H\delta \phi + (\delta H)\phi]^2 \qquad (3.29)$$

and the second variation is positive regardless of the boundary conditions applied. This means that the method seeks the solution which gives the smallest possible value of the functional. Consequently, as long as the boundary conditions applied to the trial function are consistent with those of Equation 3.1, the method of least squares will tend to the solutions of Equation 3.1. It

is not necessary to apply the extra boundary conditions (although
in Section 3.4 it will be shown that it is often advantageous to do
so). This is why the method of least squares, in the restricted
sense of Section 2.4, works. Since all that is required is that
trial functions be consistent with the boundary conditions, this
procedure can be applied to initial-value problems. Another requirement from Section 3.1 is therefore satisfied.

In the nonlinear case, it appears from Equation 3.28 that for
solutions which do not tend to satisfy Equation 3.5, a minimum
is not obtained in general, while a minimum is obtained for solutions which do tend to satisfy Equation 3.1, regardless of the
boundary conditions imposed. The conclusions about boundary
conditions reached earlier for linear problems thus carry over
to the case of nonlinear problems.

There is one essential difference between linear and nonlinear
problems. Use of a linear combination of trial functions in a linear problem leads to a unique solution for the coefficients. Thus,
as long as the restrictions on the trial functions are consistent
with Equation 3.1, we must get a good representation of ϕ. In
nonlinear problems, on the other hand, the solution for the coefficients is not unique. It is therefore possible to obtain solutions
that tend to Equation 3.6, but not to Equation 3.1, in addition to
those that tend to Equation 3.1. We, therefore, have the problem
of identifying the proper solution. It should be noted, however,
that this problem occurs in nonlinear problems even when the variational principle involved yields the proper equation of proper
order, Equation 3.1, as the Euler equation. When Equation 3.1
is nonlinear, it generally has multiple solutions. Since we already have the problem of identifying the proper solution from
among the various solutions of Equation 3.1, the least-squares
approach does not introduce an additional difficulty.

Another view of the matter of boundary conditions is obtained
by noting that the functional Equation 3.5 takes on its minimum
value of zero if, and only if, Equation 3.1 is satisfied, as long
as the restrictions placed on the trial functions are consistent
with Equation 3.1.

The same versatility with respect to boundary conditions is obtained in constrained problems. Varying Equation 3.13, we have
the second variation

$$\frac{1}{2} \delta^2 J = \int_a^b dx \left[(\delta \lambda) \frac{\partial}{\partial \lambda} (H\phi - f) + H\delta\phi + \delta_\phi H)\phi \right]^2$$

$$+ \int_a^b dx \, (\delta g)^2 + (\delta \lambda)^2 \int_a^b dx \, (H\phi - f) \frac{\partial^2}{\partial \lambda^2} (H\phi - f)$$

$$+ \int_a^b dx \, \delta^2 g \left[\int_a^b dx \, g - K \right]$$

$$+ 2(\delta \lambda) \int_a^b dx \, (H\phi - f) \frac{\partial}{\partial \lambda} [H\delta\phi + (\delta_\phi H)\phi]$$

$$+ \int_a^b dx \, (H\phi - f)[\delta_\phi^2 H)\phi + 2(\delta_\phi H)\delta\phi] \qquad (3.30)$$

If the equation and the constraint are linear (note that in this representation λ is treated as a function, so a term such as $\lambda\phi$ is a nonlinearity), Equation 3.30 reduces to the positive-definite quantity

$$\int_a^b dx \left[(\delta\lambda) \frac{\partial}{\partial\lambda} (H\phi - f) + H\delta\phi\right]^2 + \int_a^b dx \, (\delta g)^2$$

$$(3.31)$$

The arguments and conclusions are similar to those of the unconstrained case. The conclusions for the local constraint Equation 3.19 are also similar.

Let us consider, as an illustrative example of how the method of least squares can be used without extra boundary conditions, the nonlinear boundary-value problem

$$y \frac{d^2y}{dx^2} + \left(\frac{dy}{dx}\right)^2 = 1 \qquad (3.32)$$

with the conditions

$$\frac{dy}{dx}(0) = 0 \qquad (3.33)$$

$$y(1) = \sqrt{2} \qquad (3.34)$$

The exact solution is

$$y_{exact}(x) = \sqrt{1 + x^2} \qquad (3.35)$$

A simple polynomial trial solution satisfying the boundary conditions is

$$y_{approx}(x) = \sqrt{2}\,x^2 + A(1 - x^2) \tag{3.36}$$

Use of Equation 3.36 in the functional Equation 3.5 with $p(x) = 1$ yields $A = 1.083$. The exact and approximate solutions are compared in Table 3.1. It is seen that the maximum error is only 8 percent.

Table 3.1. Comparison of y_{exact} and y_{approx}

x	0	0.2	0.4	0.6	0.8	1.0
y_{exact}	1.000	1.020	1.077	1.166	1.261	1.414
y_{approx}	1.083	1.096	1.136	1.202	1.295	1.414

3.4 The Application of Extraneous Boundary Conditions

The extraneous boundary conditions may be used to increase the number of trial functions without increasing the number of trial parameters to be determined. Suppose, for example, that we have the expansion

$$\phi \approx \sum_{i=1}^{N} a_i \phi_i \tag{3.37}$$

where ϕ_i and a_i are trial functions and unknown parameters, respectively. Suppose further that we wish to include an additional trial term $bz(x)$, but that we wish to avoid the extra computing time required by the enlarged resultant system of algebraic equations. We could then specify b in terms of a_i such that one of the extraneous boundary conditions is satisfied. A significant improvement in accuracy over the expansion Equation 3.37 may result.

As a simple example, consider the equation

$$\frac{dy}{dt} + y = 0 \qquad y(0) = 1 \tag{3.38}$$

Let us examine two cases in the t interval $(0, 0.5)$,

$$y_1 = 1 + At \tag{3.39}$$

$$y_2 = 1 + A_1 t + A_2 t^2 \tag{3.40}$$

In Equation 3.40, A_1 is the arbitrary parameter and A_2 is determined in terms of A_1 such that Equation 3.38 is satisfied at $t = 0.5$, the extraneous boundary condition. Thus, y_1 and y_2 have the same number of arbitrary parameters (one), but y_2 has one more trial function. Use of the functional Equation 3.5 with $p(x) = 1$ yields the results in Table 3.2, with y_{exact} the exact solution

$$y_{exact} = e^{-t} \tag{3.41}$$

It is clear that y_2 is a significant improvement over y_1.

Table 3.2. Comparison of y_{exact}, y_1, and y_2

t	0	0.1	0.2	0.3	0.4	0.5
y_{exact}	1.000	0.905	0.819	0.741	0.670	0.607
y_1	1.000	0.914	0.828	0.742	0.656	0.570
y_2	1.000	0.905	0.818	0.738	0.666	0.602

3.5 Eigenvalue Problems

In Chapter 2 eigenvalues were obtained by substituting trial functions in the homogeneous equation, setting up the resulting weighted-residual algebraic equations, and setting the determinant of this system equal to zero. This procedure was found to be inadequate for general weighted-residual techniques other than conventional variational methods.

In this section we propose that eigenvalue problems be considered constrained problems, the constraint being the normalization condition. A functional such as Equation 3.12 is then varied with respect to the function and with respect to the eigenvalue to yield a set of algebraic equations. Because the functional contains the constraint, these algebraic equations are inhomogeneous. A key difference, then, between the least-squares variational method and more conventional weighted-residual techniques, is that, in this new approach, the eigenvalue occurs as a trial parameter in the same sense as the trial parameters for the eigenfunction occur. A secular equation (setting a determinant equal to zero) does not arise.

Let us first consider the self-adjoint equation

$$\frac{d^2 y}{dx^2} + k^2 y = 0 \qquad \frac{dy}{dx}(0) = y(1) = 0 \tag{3.42}$$

with a general normalization condition

$$\int_0^1 dx\, g(y) = 1 \tag{3.43}$$

A suitable functional could be

$$I = \int_0^1 \left(\frac{d^2y}{dx^2} + k^2 y\right)^2 dx + c^2\left(\int_0^1 dx\, g - 1\right)^2 \tag{3.44}$$

Setting the variation with respect to k^2 equal to zero, we have

$$\int_0^1 dx\, y\left(\frac{d^2y}{dx^2} + k^2 y\right) = 0 \tag{3.45}$$

If the trial solution is

$$y \approx Af(x) \tag{3.46}$$

then Equation 3.45 yields the following expression for the eigenvalue k^2:

$$\int_0^1 dx\, f(x)\left(\frac{d^2f}{dx^2} + k^2 f\right) = 0 \tag{3.47}$$

Equation 3.47 is exactly the same as the expression for k^2 which is obtained with the same trial solution from the self-adjoint variational principle, Equation 2.4.

When the trial solution Equation 3.46 is used, the equation for k^2 is independent of the normalization condition Equation 3.43. If additional trial functions are introduced, for example, if

$$y \approx A_1 f_1(x) + A_2 f_2(x) \tag{3.48}$$

this independence is no longer present. (In the conventional variational procedure, it is present regardless of the number of trial functions.) The weak coupling (weak in the sense that such coupling does not appear unless at least two trial functions are used) appears because the squared residual, being a nonlinear quantity, is affected by the magnitude of, as well as by the shape of the eigenfunction.

The least-squares variational approach to eigenvalue problems may prove to be of great significance because of its ability to treat non-self-adjoint eigenvalue problems without introducing adjoint functions. Consider, for example, the problem

$$\frac{d^3y}{dx^3} - \lambda^3 y = 0 \quad y(0) = \frac{dy}{dx}(0) = \frac{dy}{dx}(1) = 0 \qquad (3.49)$$

with a general constraint such as Equation 3.43. The exact solution is

$$y \propto x^2 e^{-\lambda x} \quad \lambda = -2 \quad \lambda^3 = -8 \qquad (3.50)$$

Two functionals will be considered

$$I_1 = \int_0^1 dx \; x^2 \left(\frac{d^3y}{dx^3} - \lambda^3 y\right)^2 + c^2 \left(\int_0^1 g \; dx - 1\right)^2$$

$$(3.51)$$

$$I_2 = \int_0^1 dx \left(\frac{d^3y}{dx^2} - \lambda^3 y\right)^2 + c^2 \left(\int_0^1 g \; dx - 1\right)^2 \qquad (3.52)$$

The reason for the x^2 weighting in Equation 3.51 is that the boundary conditions reveal more information in the vicinity of $x = 0$ than near $x = 1$. Consequently, the functional is weighted so as to emphasize errors near $x = 1$. Variations with respect to λ in Equations 3.51 and 3.52 yield, respectively, the equations

$$\int_0^1 dx \; x^2 y \left(\frac{d^3y}{dx^3} - \lambda^3 y\right) = 0 \qquad (3.53)$$

$$\int_0^1 dx \; y \left(\frac{d^3y}{dx^3} - \lambda^3 y\right) = 0 \qquad (3.54)$$

A simple polynomial trial function satisfying the boundary conditions is

$$y \approx a\left(x^2 - \frac{2}{3}x^3\right) \qquad (3.55)$$

Use of Equation 3.55 in Equation 3.53 yields

$$\lambda = -2.27 \quad \lambda^3 = -11.7 \qquad (3.56)$$

while Equation 3.54 yields

$$\lambda = -1.96 \quad \lambda^3 = -7.47 \qquad (3.57)$$

Comparing these results with the exact solution Equation 3.50, we see that the x^2 weighting improves the answers considerably. In Equation 3.57 both λ and λ^3 (the eigenvalue as it appears in Equation 3.49) are well represented, but in Equation 3.56, even though the value of λ is fairly accurate (about 13.5 percent error), the value of λ^3 is not very accurate. The reason for this is probably that residuals are much more sensitive to λ than to λ^3 because λ, and not λ^3, determines the relaxation length in Equation 3.50.

With a rather elementary trial function, Equation 3.55, and using the functional Equation 3.51, we were able to obtain accuracy to 4 percent in λ and 7 percent in λ^3. It should be clear that least-squares variational methods may prove to be very useful in the estimation of eigenvalues of non-self-adjoint equations.

3.6 The Problem of Relative Weight

3.6.1 The Problem. When faced with a system of equations, we may use a functional that is a linear combination of square residuals. If $\langle a, b \rangle$ denotes the scalar product of the vector a with the vector b and if c is a diagonal matrix with no negative terms, then a general functional is

$$I = \int_a^b dx\, p(x) \langle (H\phi - f), c(H\phi - f) \rangle \tag{3.58}$$

where H is a matrix operator, ϕ an unknown vector, f a known vector, and $p(x)$ a positive weighting function. The elements of c determine the relative importance of each of the equations.

In "ordinary" systems of equations in which all variables appear in each equation, it is logical to weight the equations equally, that is, to let c be the unit matrix, assuming that the terms in each equations are of the same order. Consider, for example, the equations

$$\frac{dy}{dt} + z = 0 \qquad y(0) = 1 \tag{3.59}$$

$$\frac{dz}{dt} - y = 0 \qquad z(0) = 0 \tag{3.60}$$

In matrix form the equations become

$$\begin{bmatrix} \dfrac{d}{dt} & +1 \\[2ex] -1 & \dfrac{d}{dt} \end{bmatrix} \begin{bmatrix} y \\[2ex] z \end{bmatrix} = \begin{bmatrix} 0 \\[2ex] 0 \end{bmatrix} \tag{3.61}$$

The exact solutions are

$$y_{exact} = \cos t \tag{3.62}$$

$$z_{exact} = \sin t \tag{3.63}$$

For t in the range $(0, 1)$, we consider the trial functions

$$y_{approx} = 1 + A_1 t^2 \tag{3.64}$$

$$z_{approx} = A_2 t \tag{3.65}$$

The functional Equation 3.58 is used with $\underset{\sim}{c}$ a unit matrix and $p(x)$ unity. The answers are given in Table 3.3. The simple trial functions yield fairly good results.

Table 3.3. Comparison of y_{approx}, z_{approx} with y_{exact}, z_{exact}

t	0	0.2	0.4	0.6	0.8	1.0
y_{approx}	1.000	0.920	0.839	0.759	0.678	0.598
y_{exact}	1.000	0.980	0.921	0.825	0.697	0.540
z_{approx}	0.000	0.170	0.340	0.511	0.681	0.851
z_{exact}	0.000	0.199	0.384	0.565	0.717	0.842

Thus, in systems such as Equation 3.58, the matrix $\underset{\sim}{c}$ serves mainly to keep dimensions and orders of magnitude consistent among the several equations. If one of the equations is a constraint however, the situation is somewhat different.

Let us illustrate this difference with an example. The problem of neutron diffusion in the presence of xenon poisoning is governed by the equations

$$D \frac{d^2\phi}{dx^2} + \Sigma\phi - \frac{y\Sigma_f}{\phi + \phi_x} \phi^2 = 0 \tag{3.66}$$

$$\int_a^b dx\, \phi = P \tag{3.67}$$

The constraint Equation 3.67 determines the power level of the nuclear reactor; y, Σ_f, ϕ_x, and D are constants; and Σ is the control variable.

By writing Equation 3.66 in convenient form, we choose a functional with the form

$$\int_a^b dx \left[(\phi + \phi_x)D \frac{d^2\phi}{dx^2} + \Sigma\phi(\phi + \phi_x) - y\Sigma_f\phi^2\right]^2 + k^2\left(\int_a^b dx\, \phi - P\right)^2$$

$$(3.68)$$

Variation of Σ yields

$$\int_a^b dx\, (\phi + \phi_x)\phi\left[(\phi + \phi_x)D \frac{d^2\phi}{dx^2} + \Sigma\phi(\phi + \phi_x) - y\Sigma_f\phi^2\right] \qquad (3.69)$$

Note that Equation 3.69 does not involve the constraint Equation 3.67. The variation with respect to ϕ, on the other hand, involves both Equation 3.66 and the constraint. Thus, the mechanics of the process tends to favor the equation at the expense of the constraint. As a consequence, the weighting constant k^2 must be used to increase the emphasis given to the constraint. The problem of relative weight is the choice of an appropriate k^2.

If we choose to use only one trial function, that is,

$$\phi \approx Au(x) \qquad (3.70)$$

then one simple solution to the problem is to make k^2 very large, in fact effectively infinite. Equation 3.69 is not affected, since it does not involve k^2. The ϕ-variation of Equation 3.68, however, becomes a statement that the constraint Equation 3.67 be satisfied exactly.

Let us now consider a numerical example using the set of numbers

$$\frac{d\phi}{dx}(0) = 0$$

$$\phi(a) = 0$$

$$a = 43.11 \text{ cm}$$

$$\Sigma_f = 0.1 \text{ cm}^{-1}$$

$$D = 1 \text{ cm}$$

$$(3.71)$$

and choosing the power constraint such that

$$\phi(0) = 10 \phi_x \tag{3.72}$$

The variational approximation will be compared with the solution of the approximate equation

$$D\frac{d^2\phi}{dx^2} + \Sigma\phi - y\Sigma_f\phi = -y\Sigma_f\phi_x \tag{3.73}$$

obtained by expanding the $\phi^2/(\phi + \phi_x)$ term. Equation 3.73 yields

$$\phi_1 = 6 \times 10^{13}(2 \cos 0.02429x - 1) \tag{3.74}$$

$$\Sigma_1 = 0.00649 \text{ cm}^{-1} \tag{3.75}$$

With

$$u(x) = \cos\frac{\pi x}{2a} \tag{3.76}$$

in Equation 3.70, the variational method with k^2 infinite yields

$$\phi_2 = 6.165 \cos\frac{\pi x}{2a} \tag{3.77}$$

$$\Sigma_2 = 0.00664 \text{ cm}^{-1} \tag{3.78}$$

The functions ϕ_1 and ϕ_2 are compared in Table 3.4.

Table 3.4. Comparison of ϕ_1 and ϕ_2

x/a	0	0.2	0.4	0.6	0.8	1.0
$10^{-13}\phi_1$	6.000	3.740	4.963	3.710	2.018	0
$10^{-13}\phi_2$	6.165	5.863	4.987	3.624	1.905	0

If several trial functions are used, however, an infinite value
of k^2 causes several equations to become degenerate. In the preceding example, for instance, each of the variational equations,
except Equation 3.69, is a statement that the constraint be satis-

fied exactly. It then becomes necessary to use a finite k^2 and to face the problem of how to choose this value.

As the number of trial functions is increased, the problem of selection should tend to disappear; because, even though the constraint equation is involved in one less variational equation than is the system equation, the total number of equations is large, and the consequences may not be severe. The case of two trial functions should pose the greatest difficulty. Several examples will be discussed in Section 3.6.2. Then, in Section 3.6.3, some general approaches to the problem will be discussed.

3.6.2 Some Examples. The effect of relative weighting will be illustrated in several examples. The first example compares the results with unit and infinite weighting when a single trial function is involved. The second concerns the use of two trial functions, in which case infinite weighting cannot be used. The third considers simultaneous constrained equations. The equations selected for these examples are chosen on the basis of simplicity and computational convenience.

Consider the equation

$$\frac{dy}{dt} + y = \lambda \qquad y(0) = 2 \tag{3.79}$$

where λ is a control parameter determined by the constraint

$$\int_0^1 y(t)\, dt = 2 - e^{-1} = 1.632 \tag{3.80}$$

The exact solution is

$$y_{exact} = 1 + e^{-t} \qquad \lambda = 1 \tag{3.81}$$

The functional to be used is

$$I = \int_0^1 dt \left(\frac{dy}{dt} + y - \lambda\right)^2 + k^2 \left(\int_0^1 y\, dt - 1.632\right)^2 \tag{3.82}$$

The trial function considered is

$$y_{approx} = 2 + At \tag{3.83}$$

Two values of k^2 were investigated, one and infinity. The control parameters obtained are for $k^2 = 1$

$$\lambda = 1.172 \tag{3.84}$$

for a 17 percent error, and for $k^2 = \infty$

$$\lambda = 0.896 \tag{3.85}$$

for a 10 percent error. The representation of the function is also better for infinite k^2, as can be seen from Table 3.5.

Table 3.5. Comparison of y_{exact}, $y_{approx}(k^2 = \infty)$, and $y_{approx}(k^2 = \infty)$ for Equation 3.79

t	0	0.2	0.4	0.6	0.8	1.0
$y_{approx}(k^2 = 1)$	2.0	1.890	1.779	1.669	1.558	1.449
$y_{approx}(k^2 = \infty)$	2.0	1.853	1.706	1.558	1.411	1.264
y_{exact}	2.0	1.819	1.670	1.549	1.449	1.368

Let us now consider Equation 3.79 when it is approximated by two trial functions

$$y_2 = 2 + A_1 t + A_2 t^2 \tag{3.86}$$

The weighting constant k^2 should be large but finite. When $k^2 = 1$ is used, the poor result $\lambda = 1.347$ is obtained. When $k^2 = 10$ is used, however, the control parameter is

$$\lambda = 0.991 \tag{3.87}$$

with less than a one percent error. The function, with $A_1 = 0.915$ and $A_2 = 0.273$, is given in Table 3.6.

Table 3.6. Comparison of y_{exact} and $y_2(k^2 = 10)$ for Equation 3.79

t	0	0.2	0.4	0.6	0.8	1.0
y_2	2.0	1.828	1.678	1.550	1.443	1.358
y_{exact}	2.0	1.819	1.670	1.549	1.449	1.368

Next, consider the simultaneous equations

$$\frac{dy}{dt} + z = \lambda \qquad y(0) = 1 \tag{3.88}$$

$$\frac{dz}{dt} - y = 0 \qquad z(0) = 1 \tag{3.89}$$

with λ determined by the constraint

$$\int_0^{0.5} dt\,(y + z) = 1.5 + \sin 0.5 - \cos 0.5 = 1.108 \tag{3.90}$$

The exact solution is

$$\left. \begin{aligned} y &= \cos t \\ z &= 1 + \sin t \\ \lambda &= 1 \end{aligned} \right\} \tag{3.91}$$

Consider the functional

$$I = \int_0^{0.5} \left(\frac{dy}{dt} + z - \lambda\right)^2 dt + k_1^2 \int_0^{0.5} dt \left(\frac{dz}{dt} - y\right)^2$$

$$+ k_2^2 \left[\int_0^{0.5} dt\,(y + z) - 1.108\right]^2 \tag{3.92}$$

Since the second and third terms contain the same variables, they may be weighted equally with respect to the first term, that is, we may let

$$k^2 = k_1^2 = k_2^2 \tag{3.93}$$

The trial functions considered are

$$\left. \begin{aligned} y_{\text{approx}} &= 1 + At^2 \\ z_{\text{approx}} &= 1 + Bt \end{aligned} \right\} \tag{3.94}$$

Since only one trial parameter is introduced for each variable, the use of infinite k^2 is permitted. Three values of k^2 are considered, 1, 10, and ∞. The values of λ obtained are

$$\left. \begin{array}{ll} \lambda = 0.889 & (k^2 = 1) \\[1em] \lambda = 1.030 & (k^2 = 10) \\[1em] \lambda = 1.137 & (k^2 = \infty) \end{array} \right\} \qquad (3.95)$$

As k^2 is increased from unity, the results get better and then get worse; the percent errors for unit and infinite k^2 are both about 11 percent. This may be explained in terms of one equation being de-emphasized with respect to the others, that is, in terms of overcompensation by making k^2 too large. In Table 3.7, the approximate solutions for when k^2 is ten and infinity are compared with the exact solutions. It is seen that for $k^2 = \infty$ the results for z are significantly better than those for y, while for $k^2 = 10$, the accuracy is more equitably distributed between y and z. In addition, the constraint is well satisfied for $k^2 = \infty$.

$$\int_0^{0.5} (y_{ap} + z_{ap})dt = 1.113 \qquad (3.96)$$

Equation 3.88 thus tends to be de-emphasized when very large values of k^2 are used. This phenomenon must be taken into account when analyzing simultaneous constrained equations.

Table 3.7 Comparison of Exact and Approximate Solutions to Equations 3.88 and 3.89

t	0	0.1	0.2	0.3	0.4	0.5
$y_{approx}(k^2 = \infty)$	1.000	0.998	0.991	0.981	0.965	0.946
$y_{approx}(k^2 = 10)$	1.000	0.996	0.984	0.963	0.935	0.898
y_{exact}	1.000	0.995	0.980	0.955	0.921	0.878
$z_{approx}(k^2 = \infty)$	1.000	1.098	1.196	1.294	1.392	1.490
$z_{approx}(k^2 = 10)$	1.000	1.094	1.188	1.282	1.376	1.470
z_{exact}	1.000	1.100	1.199	1.296	1.389	1.479

3.6.3 General Approaches to the Problem. It is desirable to have available methods of selecting a value of k^2 that will lead to good results. Two approaches will be considered. One is to force all the variational equations to contain all of the system equations. The second is to force the system and constraint equations to have equal residuals, after making dimensional corrections.

Let us illustrate these two approaches with reference to the system of equations

$$H\phi - f = 0 \tag{3.97a}$$

$$\int_a^b dx\, g(\phi) - c = 0 \tag{3.97b}$$

where, in general, $H = H(\lambda)$ and $f = f(\lambda)$. Consider the functional

$$I = \int_a^b dx\, (H\phi - f)^2 + k^2 \left(\int_a^b dx\, g - c \right)^2 \tag{3.98}$$

In the first approach, we may let

$$k^2 = p(\lambda) \tag{3.99}$$

where $p(\lambda)$ is any positive function of λ. The variation with respect to λ then yields

$$0 = \int_a^b dx\, (H\phi - f)\, \frac{\partial}{\partial \lambda}\, (H\phi - f) + p\, \frac{dp}{d\lambda} \left(\int_a^b dx\, g - c \right)^2 \tag{3.100}$$

Use of Equation 3.99 thus results in the constraint Equation 3.97b being represented in the λ variation.

The second approach assumes that if all equations are equally in error, then the results for all variables should approximately be of equal accuracy. We therefore go through the variational procedure, deriving a set of algebraic equations, while leaving k^2 undetermined, and then we impose a condition of the form

$$\int_a^b dx\, (H\phi - f)^2 = \left(\int_a^b dx\, g - c \right)^2 \tag{3.101}$$

Each of these methods has the disadvantage that the resulting equations are more strongly nonlinear than if a number were intuitively substituted for k^2. In addition, the second method increases the number of parameters to be determined. It is also possible that the value of k^2 obtained by invoking Equation 3.101 could be negative, thus removing the positive-definite property of the functional.

The higher degree of nonlinearity is not necessarily a serious problem. If the equations are already sufficiently nonlinear that a gradient method or an iterative procedure is necessary to solve the variational equations, then the added nonlinearity may be acceptable. The analyst must judge for himself what approach is best for his particular problem.

3.7 Comparison with Conventional Variational Methods

3.7.1 Fourier Expansions and Linear Inhomogeneous Equations.
When the solution of a linear inhomogeneous equation is expanded in the orthogonal or biorthogonal eigenfunctions of the homogeneous operator, a Fourier series results, with the attendant implications about minimum-square errors or stationary joint errors. The orthogonality and biorthogonality relations of the homogeneous modes and the error interpretations of the Fourier expansions are summarized for the convenience of the reader in Appendix D.

Consider an equation of the form

$$H\phi - \lambda M\phi - f(x) = 0 \qquad (3.102)$$

Assume that H and M are self-adjoint operators. Let the solution be expanded in terms of the eigenfunctions of the homogeneous equation

$$H\psi_n = \lambda_n M\psi_n \qquad (3.102')$$

$$\phi = \sum_{n=1}^{N} a_n \psi_n \qquad (3.103)$$

The self-sufficient functional approach yields the set of equations

$$\int_a^b dx\, \psi_m \left[\sum_{n=1}^{N} a_n (H - \lambda M)\psi_n - f \right] = 0 \qquad (3.104)$$

for $m = 1, \cdots, N$. Invoking Equation 3.102' and the orthogonality condition

$$\int_a^b dx\, \psi_n M\psi_m = \delta_{nm} \qquad (3.104')$$

(see Appendix D, Equation D.6) we have

$$a_n = \frac{b_n}{\lambda_n - \lambda} \tag{3.105}$$

$$b_n = \int_a^b dx\, f(x)\, \psi_n(x) \tag{3.106}$$

The results, Equations 3.105 and 3.106, are just the Fourier co-efficients for ϕ. In other words, the use of the modes of the homogeneous problem as trial functions in the self-sufficient functional yields a Fourier series for the solution. This variational method therefore implies (for these trial functions) the error-minimizing properties of the Fourier series discussed in Appendix D.
     If the least-squares functional is

$$I = \int_a^b dx\, p(x)(H\phi - \lambda M\phi - f)^2 \tag{3.107}$$

then the first variations yield

$$\int_a^b dx\, p(x)(\lambda_m - \lambda)(M\psi_m) \sum_{n=1}^{N} (\lambda_n - \lambda)a_n M\psi_n - f = 0 \tag{3.108}$$

for $m = 1, \cdots, N$.
     In the case

$$M = q(x) \tag{3.109}$$

$$p(x) = \frac{1}{q(x)} \tag{3.110}$$

the least-squares approach also yields the Fourier coefficients, Equations 3.105 and 3.106. Thus, for a large class of problems, appropriate choice of the weighting factor $p(x)$ in the least-squares approach leads to the same results as given by the self-sufficient functional approach.
     If M is a more general operator or if a different weighting is chosen for the square residual, the results may still be approximately the same as for the self-sufficient functional approach. Assume $f(x)$ to be well represented by the expansion

$$f(x) = \sum_{n=1}^{N} b_n M \psi_n \qquad (3.111)$$

$$b_n = \int_a^b dx\, f(x)\, \psi_n \qquad (3.112)$$

Substitution into Equation 3.108 yields

$$\int_a^b dx\, p(x)(M\psi_m) \sum_{n=1}^{N} [(\lambda_n - \lambda)a_n - b_n] M\psi_n = 0$$

$$(3.113)$$

for $m = 1, \cdots, N$. The solution of Equation 3.113 is obviously Equation 3.105. The criterion of validity of this solution is that Equation 3.111 be a good representation of the inhomogeneous term. This is also the criterion of validity of the Fourier expansion of N terms. In other words, both methods tend to give the same results, provided good results are possible.

If the operators H and M are not self-adjoint, a dual functional yields, analogous to Equation 3.104,

$$\int_a^b dx\, \psi_m^* \left[\sum_{n=1}^{N} a_n(\lambda_n - \lambda)M\psi_n - f \right] \qquad (3.114)$$

for $m = 1, \cdots, N$ and the coefficients are

$$a_n = \frac{b_n}{\lambda_n - \lambda} \qquad (3.115)$$

$$b_n = \int_a^b dx\, f\psi_n^* \qquad (3.116)$$

which are the Fourier coefficients for the biorthogonal series.

The least-squares approach again yields Equation 3.108. Let us again assume that we may make the expansion Equation 3.111 with b_n in this case given by Equation 3.116. This leads to Equation 3.113 again, so the coefficients a_n are given by Equation 3.115. Here too, the criterion for the assumption Equation 3.111 is the same as for the use of the Fourier expansion with N terms.

In special cases, the approximate correspondence between methods may actually be exact. Suppose, for example, that $M\psi_m$ is proportional to ψ_m. Such is the case for the equation

$$\frac{d^2}{dx^2}\left(\frac{d^2}{dx^2} + k^2\right)\psi_n = 0 \qquad (3.117)$$

with boundary conditions such that

$$\psi_n = \cos k_n x \qquad (3.118)$$

Equation 3.108 then yields Equations 3.105 and 3.106.

It has been stated that the criterion for the validity of the conventional variational results and the criterion for the least-squares approach to yield essentially the same results as conventional variational procedures are both the validity of the expansion Equation 3.111 with the coefficients given by Equations 3.112 or 3.116. We may ask, however, which method gives better results when Equations 3.111 and 3.112 constitute merely a crude representation of the inhomogeneity.

The answer, of course, depends on the criterion selected by which to judge the solution. As is shown in Appendix D, for self-adjoint operators where M is merely a function (for example, not a derivative operator), the Fourier series, and therefore the variational method, minimizes the weighted-square error in the function. If the criterion is minimum-square error with this weighting, the variational method gives the best possible answer.

There is, however, another point that should be considered: The procedure may completely ignore a significant part of the source term f and, therefore, a significant part of the solution. The consequences can be serious, particularly when an error-minimizing interpretation (see Appendix D, Section D.2) cannot be invoked. Since the method of weighted least squares makes use of all the information provided (that is, does not neglect part of the source term f) and since it has a residual-error-minimizing interpretation, we can argue that this method is more reliable than conventional variational methods when the Fourier expansion Equation 3.111 with Equation 3.112 or with Equation 3.116 introduces a significant error.

3.7.2 General Expansions. We shall consider linear and non-linear problems in this section.

Linear problems. Let us compare the results of the least-squares and conventional variational methods for the case of an arbitrary set of trial functions. The solution of Equation 3.102 is expanded in the series

$$\phi = \sum_{n=1}^{N} a_n \theta_n \qquad (3.119)$$

In the dual-functional method we also expand

$$\phi^* = \sum_{n=1}^{N} a_n^* \theta_n^* \qquad (3.120)$$

The least-squares, self-sufficient functional, and dual-functional methods yield, respectively,

$$\int_a^b dx \, p(x)[(H - \lambda M)\theta_m]\left[\sum_{n=1}^{N} a_n(H - \lambda M)\theta_n - f\right] = 0 \qquad (3.121)$$

for $m = 1, \cdots, N$;

$$\int_a^b dx \, \theta_m\left[\sum_{n=1}^{N} a_n(H - \lambda M)\theta_n - f\right] = 0 \qquad (3.122)$$

for $m = 1, \cdots, N$;

$$\int_a^b dx \, \theta_m^*\left[\sum_{n=1}^{N} a_n(H - \lambda M)\theta_n - f\right] = 0 \qquad (3.123)$$

for $m = 1, \cdots, N$; Equation 3.122 applies if H and M are self-adjoint operators. Let the set of functions y_n be the set generated by the homogeneous operator operating on the trial functions; that is,

$$y_n = (H - \lambda M)\theta_n \qquad (3.124)$$

Let the inhomogeneous term $f(x)$ be expanded in the set y_n according to some criterion (such as a least-squares fit)

$$f(x) = \sum_{n=1}^{N} b_n y_n \qquad (3.125)$$

All three approaches then yield

$$a_n = b_n \tag{3.126}$$

It is therefore to be expected that if the expansion Equation 3.125 is not made, then the various methods will yield results which are in the vicinity of Equation 3.126. Consequently, the "best" potential performance is approximately the same for each method; that is, none can do very much better than Equation 3.126. On the other hand, there is considerable variation in the "worst" potential performance. If the adjoint trial functions θ_m^* do not provide a good representation for the solution of the adjoint equation, then Equation 3.123 may yield significantly poorer results than would Equation 3.121. It thus appears that unless we are interested in the adjoint function or in the value of the functional of the dual-functional approach, the method of least squares provides a more reliable procedure for linear inhomogeneous equations than does the dual-functional variational method.

Nonlinear problems. When a general expansion is used for a nonlinear problem, the conclusions are similar to those for the linear case, except that not all of the solutions need be alike for each approach. (A nonlinear problem, in general, may have several solutions.)

Consider the equation

$$H\phi - f = 0 \tag{3.127}$$

and the expansion

$$\phi = \sum_{n=1}^{N} a_n \theta_n \tag{3.128}$$

The dual- and self-sufficient functional approaches yield

$$\int_a^b dx \; \theta_m^* \sum_{n=1}^{N} (a_n H \theta_n - f) = 0 \tag{3.129}$$

for $m = 1, \cdots, N$;

$$\int_a^b dx \; \theta_m \sum_{n=1}^{N} (a_n H \theta_n - f) = 0 \tag{3.130}$$

for $m = 1, \cdots, N$; and the least-squares approach yields

$$\int_a^b dx \; p(x) \left(H\theta_m + \sum_{n=1}^{N} a_n \frac{\partial H}{\partial a_m} \theta_n \right) \left(\sum_{n=1}^{N} a_n H\theta_n - f \right) = 0 \qquad (3.131)$$

for $m = 1, \cdots, N$. The nonlinear operator H generates the set

$$z_n(a_1, a_2, \cdots, a_N, x) = H\theta_n \qquad (3.132)$$

Let f be expanded in this set

$$f = \sum_{n=1}^{N} b_n(a_1, a_2, \cdots, a_N) z_n \qquad (3.133)$$

The expansion must be made symbolically in terms of the coefficients a_n, because the a_n are still unknown parameters. Then

$$\sum_{n=1}^{N} a_n H\theta_n - f = \sum_{n=1}^{N} (a_n - b_n) z_n \qquad (3.134)$$

and the a_n, as determined from the set of nonlinear equations

$$a_n = b_n(a_1, a_2, \cdots, a_N) \qquad (3.135)$$

for $n = 1, a, \cdots, N$, are the same for both the variational and least-squares techniques. Equation 3.135, however, does not necessarily yield all of the solutions. In nonlinear problems, therefore, not all solutions necessarily tend to be the same for each method. It should also be borne in mind that trial functions which are good representations of one solution may be poor representations of another.

3.7.3 Ease of Computation. In nonlinear problems, the least-squares approaches generally require greater effort than do the variational approaches. Compare, for example, the least-squares Equations 3.131 with the variational Equations 3.129 or 3.130. The least-squares approach requires that we set up the additional N^2 terms

$$a_n \frac{\partial H}{\partial a_m} \theta_n \qquad (3.136)$$

for $m, n = 1, 2, \cdots, N$ and to evaluate the additional $N^3 + N^2$ integrals

$$a_n a_r \int_a^b dx \left(\frac{\partial H}{\partial a_m} \theta_n \right) (H\theta_r) \, p(x) \qquad (3.137)$$

for $m, n, r = 1, 2, \cdots, N$;

$$a_n \int_a^b dx \left(\frac{\partial H}{\partial a_m} \theta_n \right) f \, p(x) \qquad (3.138)$$

for $m, n = 1, 2, \cdots, N$.

It thus appears that considerably more work may be entailed in least-squares approaches to nonlinear problems than is involved in conventional variational approaches. In addition, higher-order equations result for the parameters in the least-squares methods. Consider, for example, the self-sufficient equation

$$\frac{d^2 y}{dx^2} + y^2 = 0 \qquad (3.139)$$

which can be derived from the functional

$$J = \int_a^b dx \left[\frac{1}{2} \left(\frac{dy}{dx} \right)^2 - \frac{1}{3} y^3 \right] \qquad (3.140)$$

The use of a trial solution of the form

$$y = Au(x) \qquad (3.141)$$

in Equation 3.140 yields a quadratic equation for the parameter A. However, a least-squares functional, yields a cubic equation for A.

3.8 Summary.

The least-squares variational method satisfies all the criteria set forth in Section 3.1. It can treat any equation without requiring the use of adjoint functions. It can handle any type of equality constraint. Any boundary conditions can be treated. The value of the functional can serve as a measure of the accuracy of the result. All eigenvalue problems (self-adjoint or not) can be treated. Since the error parameter (the functional) is minimized with respect to variations in eigenvalues (and in control parameters in general constrained problems), accurate eigenvalues (and control parameters) are to be expected.

It should not be concluded, however, that because of these advantages the least-squares approach should be used for all problems and conventional variational (and other weighted residual methods) be relegated to oblivion. While the least-squares method seems to be the most suitable general approach, in specific applications (in which some specific criteria may be added to our "general" list) other methods may be preferable. In some cases, for example those discussed in Section 2.1, the need for an accurate value of the "figure of merit" may supersede the desire for an accurate positive definite error parameter.

The main advantage of the least-squares approach is indeed its generality. It is for this reason that several options, about which judgment must be used in the particular problem at hand, are left to the reader. The reader must select the positive weighting function $p(x)$ in accordance with his own opinion as to where in his system errors are most important. A decision of this type was made in the treatment of the eigenvalue problem discussed in Section 3.5. Similarly, the relevant information concerning the problem of relative weight is given in Section 3.6. It is left to the analyst, however, to select the proper approach for his own problem. Outside the realm of specific problems, it becomes difficult (and even dangerous) to attempt to lay down hard-and-fast rules of procedure.

Chapter 4

APPLICATION TO FUEL DEPLETION IN A
NUCLEAR REACTOR

In Chapter 3 the method of least squares was applied to several numerical examples of an elementary nature. The main purpose of this chapter is to demonstrate the applicability of the method to complex problems. The problem of fuel depletion in a nuclear reactor is used for this demonstration.

The depletion problem is nonlinear, constrained, an initial-value problem, and a partial differential equation with several dependent variables; it has characteristics that make application of conventional variational methods difficult. It is also an important nuclear engineering problem and is of the same form as other important problems, such as the "flux-tilt" oscillations caused by xenon poisoning and the short-time kinetics of nuclear reactors. A second purpose of this chapter is to investigate the feasibility of using the method of least squares for obtaining approximate solutions to these important problems.

A standard type of depletion calculation is made to obtain a reference solution with which to compare the results from the variational method. Variational calculations are made using the initial and final flux distributions of the conventional calculation as trial functions to test the method, rather than to test the adequacy of the trial functions. Excellent results are obtained.

Since in practical problems the "end-of-life" solutions are not known, it is necessary to generate a set of trial functions based on "beginning-of-life" conditions. Trial functions generated by a perturbation method are also found to yield excellent results.

4.1 Fuel Depletion, Xenon Oscillations, and Kinetics

In nuclear-reactor analysis there are three basic dynamic problems, distinguished from one another to a great extent by their relative time scales. Significant changes in flux and power distributions take place in fuel-depletion problems over the course of months, in xenon oscillations over many hours, and in kinetic problems over seconds or even over small fractions of a second, depending on the particular problem. Although these are different phenomena, occurring over vastly different time scales, the basic equations governing them are similar. Thus, a method shown to be applicable to the solution of one of these problems should be useful in dealing with the other two.

There is a need for a rapid but reasonably accurate method for the approximate solution of these problems. Many fuel-depletion calculations, for example, are made in the course of optimization studies, feasibility studies, and preliminary-design studies. In such investigations trends and relative values are of interest, rather than the detailed information provided by diffusion codes. Moreover, these codes, are time-consuming and, therefore, costly.

Let us now consider the mathematical formulation of these phenomena. A simple model of a depletion problem in a slab reactor of thickness 2a is governed by the system of equations

$$\frac{\partial}{\partial x}\left(D\,\frac{\partial \phi}{\partial x}\right) + (\nu\sigma_f - \sigma_a)N\phi - \Sigma_{ap}(t)\phi = 0 \tag{4.1}$$

$$\frac{\partial N}{\partial t} = -\sigma_a \phi N \tag{4.2}$$

$$\int_0^a dx\, N\phi = P' \tag{4.3}$$

This model assumes that the <u>one-group diffusion theory</u>, represented by Equation 4.1, is adequate to describe the neutron-flux distribution; that there is only one fissile isotope, denoted by N, initially present; that there is no buildup of fission products; and that a variable uniform poison with cross section $\Sigma_{ap}(t)$ is used to maintain criticality while the system operates at constant power P', specified by the constraint Equation 4.3. There is no time-derivative term in the diffusion Equation 4.1, because this term would be negligible compared with the other terms in the equation.

The inclusion of more energy groups of neutrons would result in more equations of the form of Equation 4.1. The inclusion of several fuels and of fission products would result in additional equations similar to Equation 4.2. The basic mathematical difficulties of the depletion problem are therefore represented in the simple model used here.

In the xenon-oscillation and short-time kinetics problems, some variation in the form of the equations may occur depending on the nature of the specific system and on the cause of the transient. Various models treat constant-power production (in the xenon problem), Newton's law of cooling, and no energy removal (in the short-time kinetics problem). There may be one or more characteristic temperatures, for example, fuel and moderator temperatures in a heterogeneous system, that introduce feedback effects. In none of these cases, however, is the set of equations fundamentally different in form from those of the depletion model discussed earlier (except for the absence of a constraint in some models).

We shall consider homogeneous slab reactors with Newton's law of cooling and one characteristic temperature. In the xenon problem, the equations are of the form

$$\frac{1}{v}\frac{\partial \phi}{\partial t} = \frac{\partial}{\partial x}\left(D\frac{\partial \phi}{\partial x}\right) + \Sigma\phi - \sigma_x N_x \phi \tag{4.4}$$

$$\frac{\partial N_x}{\partial t} = y_x \Sigma_f \phi - (\sigma_x \phi + \lambda_x)N_x + \lambda_I N_I \tag{4.5}$$

$$\frac{\partial N_I}{\partial t} = y_I \Sigma_f \phi - \lambda_I N_I \tag{4.6}$$

$$\frac{\partial T}{\partial t} = b_1 \phi - b_2(T - T_c) \tag{4.7}$$

where N_x and N_I are xenon-135 and iodine-135 concentrations, T is the system temperature, T_c is the coolant temperature, and b_1 and b_2 are constants of proportionality. Xenon-135, which has a high absorption cross section σ_x, is produced directly as a fission product and by means of radioactive decay of iodine-135, another fission product. When an asymmetric perturbation is introduced into the system, the flux shape changes. The resulting change in xenon concentration acts as a restoring force, thus tending to bring about spatial oscillations. When the flux tilts, power production is increased where the flux is increased, and the temperature tends to rise in these areas (and fall where the flux is decreased). This causes an additional feedback effect, because the quantities in Equations 4.4, 4.5, and 4.6 are temperature-dependent. This dependence is usually expressed in terms of temperature coefficients,[29] for example,

$$\Sigma = \Sigma_0[1 + \alpha(T - T_0)] \tag{4.8}$$

where α is the temperature coefficient and Σ_0 is the value of the cross-section term Σ at the reference temperature T_0.

In the kinetics problem, the equations are

$$\frac{1}{v}\frac{\partial \phi}{\partial t} = \frac{\partial}{\partial x}\left(D\frac{\partial \phi}{\partial x}\right) + [\nu(1 - \beta)\Sigma_f - \Sigma_a]\phi + \sum_{i=1}^{N} \lambda_i C_i \tag{4.9}$$

$$\frac{\partial C_i}{\partial t} = \nu\beta_i \Sigma_f \phi - \lambda_i C_i \tag{4.10}$$

where $i = 1, \cdots, M$, and

$$\frac{\partial T}{\partial t} = b_1 \phi - b_2 (T - T_c) \qquad (4.11)$$

where the β_i are the fractions of fission neutrons in the i-th delayed group, β is the total delayed neutron fraction, and C_i the
concentration of precursors for neutrons of the i-th delayed group.
In these equations, the temperature coefficients are assumed, as
in Equation 4.8.

4.2 The Standard Procedure

In this section we discuss the standard method of solution of
depletion problems and an approximate method which is based on
this standard approach.

In the standard method, the diffusion Equation 4.1 is solved
numerically by finite-difference methods as an eigenvalue problem at the initial time, yielding the value of Σ_{ap} needed for criticality and the initial-flux distribution. The magnitude of this
distribution is then determined from the constraint Equation 4.3.
The system is divided into several regions, in each of which the
average flux is computed from the initial-flux distribution. If
we assume a uniform flux in each region (at the average value),
the depletion Equation 4.2 is solved numerically in each region
to yield the fuel density (which will also be uniform in each
region) at the end of a small time step τ. Using this new fuel
distribution, we solve Equation 4.1 as an eigenvalue problem for
$\phi(x, \tau)$ and $\Sigma_{ap}(\tau)$. The procedure is then repeated and is carried
out for several time steps, usually until Σ_{ap} is reduced to
zero or to some other preset level. The time at which Σ_{ap} at-
tains its preset level is called the core life. The core life is
an important quantity, because it has a strong influence on fuel-
cycle costs.

The major disadvantage of the standard approach is that it is
time-consuming. Several diffusion calculations, each of which
is expensive to perform, must be made if the results are to be
accurate over the entire core life. This is because of the nature
of the procedure, that is, the introduction of errors by flux av-
eraging at each time step causes errors (in addition to errors
arising from finite-difference procedures) which tend to build up
toward the end of the core life.

Kaplan, and others,[13] discuss the use of synthesis techniques
(see Section 2.6.2) to permit reduction of the number of diffusion
calculations. They obtain N trial functions ψ_i for the flux by
using a standard depletion code, but taking very long time steps.
They then follow the depletion procedure, discussed earlier, with
small time steps, except that they solve the diffusion equation

approximately at each time step using Galerkin's method with the aforementioned trial functions, instead of accurately by a finite-difference method. At the j-th time step, for example, the flux $\phi_j(x)$ would be expanded in the series

$$\phi_j(x) = \sum_{i=1}^{N} a_{ji}\psi_i(x)$$ (4.12)

The coefficients a_{ji} are determined from the Galerkin equations

$$\int_a^b dx\ \psi_k R(x) = 0$$ (4.13)

where $k = 1, \cdots, N$, and $R(x)$ is the residual of the diffusion Equation 4.1.

4.3 The Least-Squares Variational Procedure

The functional used for the burnup problem for the time interval $(0, T)$ described by Equations 4.1, 4.2, and 4.3 is

$$I = \frac{1}{2} \int_0^T dt \int_0^a dx \left[D\frac{\partial^2 \phi}{\partial x^2} + (\nu\sigma_f - \sigma_a)N\phi - \Sigma_{ap}\phi \right]^2$$

$$+ \frac{1}{2}k^2 \left[\int_0^T dt \int_0^a dx \left(\frac{\partial N}{\partial t} + \sigma_a \phi N\right)^2 \right.$$

$$\left. + \frac{\sigma_a^2}{a} \int_0^T dt \left(\int_0^a dx\ N\phi - P' \right)^2 \right]$$ (4.14)

The constraint equation terms are weighted with the coefficient σ_a^2/a so as to make them of the same order of magnitude as the terms arising from the other equations. This coefficient constitutes a "dimensional" weighting correction, as discussed in Section 3.2. Once this correction is made, the same weighting is given to the depletion and constraint equations relative to the diffusion equation, because the control variable $\Sigma_{ap}(t)$ appears only in the diffusion equation.

It is assumed for the purpose of this analysis that we are willing to calculate the initial-flux distribution and poison level accurately. The trial solutions are taken to be of the form

$$\phi(x, t) \approx A_0 \sum_{i=1}^{J} \psi_i(x) \left(T_{io} + \sum_{m=1}^{L} T_{im} t^m \right) \qquad (4.15)$$

$$N(x, t) \approx B_0 N_0(x) \left(1 + \sum_{i=1}^{J} \psi_i(x) \sum_{m=1}^{L} \theta_{im} t^m \right) \qquad (4.16)$$

$$\Sigma_{ap}(t) = \Sigma_{ap}(0) \left(1 + \sum_{m=1}^{L} \rho_m t^m \right) \qquad (4.17)$$

In these equations A_0 and B_0 are the maximum values of the initial flux and fuel density, $N_0(x)$ is the initial-fuel distribution (with maximum value 1), and the $\psi_i(x)$ are the spatial trial functions (satisfying the boundary conditions of the diffusion equation) for the flux. In Equation 4.16 it is assumed that the trial functions for the flux are also capable of describing the fuel distribution. Since changes in the fuel distribution are proportional to the flux, the assumption should be valid. It is also assumed that the variables change slowly enough for a polynomial in time to be satisfactory. Since no short-lived fission products are considered, this assumption should also be valid.

Since the initial-flux distribution $\phi_0(x)$ is presumed known, the trial solution, Equation 4.15, should reduce to $\phi_0(x)$ when $t = 0$. This is accomplished by setting $\psi_1(x)$ equal to $\phi_0(x)$ and choosing

$$T_{io} = \begin{cases} 1 \text{ if } i = 1 \\ 0 \text{ if } i > 1 \end{cases} \qquad (4.18)$$

The trial solutions, Equations 4.15, 4.16, and 4.17, are substituted into the functional Equation 4.14, and the variations are taken with respect to the unknown parameters T_{im}, θ_{im}, and ρ_m. The integrals in the resulting equations are evaluated, leaving a set of simultaneous nonlinear algebraic equations. A computer program has been written to evaluate the integrals and to solve the resulting equations. In Section 4.6 results obtained using this program are compared to those of a standard code of the type described in Section 4.2.

The least-squares procedure has several desirable characteristics not possessed by the standard techniques or by approximate procedures based on these techniques. The least-squares procedure is defined over an entire time period T and does not require separate solutions for intermediate times in a step-by-step manner.

There is consequently a tendency for errors to be distributed over the period T, rather than to build up cumulatively toward the end of this period. This state of affairs is because the nonlinear problem is treated nonlinearly rather than as a sequence of linear problems. In addition, the error (square residual) is minimized with respect to variations in each of the variables of the problem: flux, fuel, and poison.

4.4 The Choice of Trial Functions

In the case to be considered in Section 4.6, the Expansions 4.15, 4.16, and 4.17 are truncated with J = 2 and with L = 1. There are consequently five unknown parameters to be determined, namely T_{11}, T_{21}, θ_{11}, θ_{21}, and ρ_1. It was noted in Section 4.3 that the first spatial trial function $\psi_1(x)$ is taken to be the initial-flux distribution $\phi_0(x)$. The second spatial trial function $\psi_2(x)$ remains to be selected.

One set of calculations is to be made using a trial function $\psi_2(x)$ which is known to be "good," so as to obtain a check on the method rather than on the suitability of the trial function. This good trial function is the flux distribution at time T, the end of the time period under consideration, as given by the depletion computer code FEVER.[28] This end-of-life trial function will hereafter be referred to as the "FEVER trial" and will be denoted by $\phi_F(x)$.

It is also desirable to perform an analysis using a trial function that does not require a prior knowledge of a detailed solution of the problem. A second variational calculation is to be made using a "perturbation trial" $\phi_P(x)$ for the second trial function $\psi_2(x)$. The perturbation technique used to generate $\phi_P(x)$ is described in Section 4.5.

4.5 Perturbation Methods for Trial Functions and Trial Parameters

We have two reasons for investigating the use of perturbation techniques in connection with the depletion problem. One reason is to generate trial functions, as was pointed out in Section 4.4. The second is to generate reasonable estimates of the quantities T_{im}, θ_{im}, and ρ_m of the trial solutions, Equations 4.15, 4.16, and 4.17, in order to facilitate the solution of the nonlinear algebraic equations determining these quantities. The numerical solution of such equations usually entails an iterative, gradient, or direct search procedure, each of which requires an initial estimate of the solution; the better the estimate, the better the prospects of rapid convergence to the proper solution.

At the beginning of core life, the fuel distribution is given approximately by

$$N(x, t) \approx N_0(x)\, B_0[1 - \sigma_a A_0 t \phi_0(x)] \qquad (4.19)$$

The use of Expression 4.19 in the diffusion Equation 4.1 yields

$$\frac{\partial}{\partial x}\left(D\,\frac{\partial\phi}{\partial x}\right) + [(\nu\sigma_f - \sigma_a)\,B_0 N_0(x) - \Sigma_{ap}(t)]\phi$$

$$- (\nu\sigma_f - \sigma_a)\,B_0 N_0(x)\,\sigma_a A_0 t\phi_0(x)\,\phi = 0$$

$$(4.20)$$

For simplicity we shall consider the case in which D and N_0 are spatially constant. Equation 4.20 can then be written in the form

$$\frac{d^2\phi}{dx^2} + \mu^2\phi - \lambda\phi_0(x)\,\phi = 0 \tag{4.21}$$

with the eigenvalue μ^2 given by

$$\mu^2 = \frac{1}{D}[(\nu\sigma_f - \sigma_a)\,B_0 N_0 - \Sigma_{ap}(t)] \tag{4.22}$$

and the perturbation term λ given by

$$\lambda = \frac{1}{D}(\nu\sigma_f - \sigma_a)\,B_0 N_0 \sigma_a A_0 t \tag{4.23}$$

Integration of Equation 4.21 twice in succession and application of the boundary conditions

$$\frac{d\phi}{dx}(0) = \phi(a) = 0 \tag{4.24}$$

yield the integral equation

$$\phi(x) = \mu^2 \int_x^a dx \int_0^x dx_2\,\phi(x_2) - \lambda \int_x^a dx_1 \int_0^{x_1} dx_2\,\phi_0(x_2)\,\phi(x_2)$$

$$(4.25)$$

or an equivalent form

$$\phi(x) = (a - x)\int_0^x dx_1\,[\mu^2 - \lambda\phi_0(x_1)]\,\phi(x_1)$$

$$+ \int_x^a dx_1\,(a - x_1)[\mu^2 - \lambda\phi_0(x_1)]\,\phi(x_1) \tag{4.26}$$

Equations 4.25 and 4.26 are exact. A first estimate of $\phi(x)$ may be obtained by use of $\phi(x) = \phi_0(x)$ in the integrands to yield

$$\phi(x) \approx \phi_0(x) - \lambda \int_x^a dx_1 \int_0^{x_1} dx_2 \; \phi_0^2(x_2) \qquad (4.27)$$

and

$$\phi(x) \approx (a - x) \int_0^x dx_1 \; [B^2 - \lambda\phi_0(x_1)] \; \phi_0(x_1)$$

$$+ \int_x^a dx_1 \; (a - x_1)[B^2 - \lambda\phi_0(x_1)] \; \phi_0(x_1) \qquad (4.28)$$

where B^2 is the unperturbed eigenvalue

$$B^2 = \mu^2(\lambda = 0) = \left(\frac{\pi}{2a}\right)^2 \qquad (4.29)$$

and $\phi_0(x)$ is given by

$$\phi_0(x) = \cos\frac{\pi x}{2a} \qquad (4.30)$$

Equation 4.27 becomes

$$\phi(x) \approx \cos\frac{\pi x}{2a} - \frac{\lambda a^2}{2\pi^2}\left\{\frac{\pi^2}{2}\left[1 - \left(\frac{x}{a}\right)^2\right] + 1 + \cos\frac{\pi x}{a}\right\}$$
$$(4.31)$$

Expression 4.31 will be used in Section 4.6 for the perturbation trial $\phi_P(x)$ chosen as the trial function $\psi_2(x)$ in Equation 4.15.

Equation 4.27 involves two successive integrations of the same quantity, while Equation 4.28 involves two separate integrations. The total amount of work is about the same in each case. Since one integration should take about the same time as one iteration of a finite-difference solution procedure for Equation 4.21, the use of the perturbation technique may result in a considerable saving of time in the generation of trial functions compared with Kaplan's method of selecting trial functions (as discussed in Section 4.2) if the perturbation-generated trial function proves adequate to describe the solution. Note that this statement does not imply that Equation 4.31 need give a good representation of $\phi(x)$ with the correct value of λ. All that is required is that the <u>form</u> of the correction term be adequate to describe the actual correction if the magnitude of the coefficient λ is allowed to vary. In Section 4.6, where numerical results are presented, this is shown to be the case for the simple depletion problem under discussion.

When fuel is depleted, the flux level must increase (in addition to the flux distribution tending to "flatten") in order for constant power to be maintained. To get an estimate of this effect, we allow A_0 to vary with time

$$A_0 \rightarrow A_{00} + A_{01}t \qquad (4.32)$$

and we invoke the constant-power constraint in the approximate form

$$\int_0^a dx\, B_0 N_0[1 - c_1 t \phi_0(x)](A_{00} + A_{01}t)[\phi_0(x) - c_2 t \phi_1(x)] = P' \qquad (4.33)$$

where

$$c_1 = \sigma_a A_0 t \qquad (4.34)$$

$$c_2 = \frac{\lambda a^2}{2\pi^2} \qquad (4.35)$$

$$\phi_1(x) = \frac{\pi^2}{2}\left[1 - \left(\frac{x}{a}\right)^2\right] + 1 + \cos\frac{\pi x}{a} \qquad (4.36)$$

as obtained from the Expressions 4.19 and 4.31. An estimate of A_{01} may be obtained by neglecting the t^2 and t^3 terms in Equation 4.33.

To estimate the change in the poison level $\Sigma_{ap}(t)$, we may obtain the change in the eigenvalue μ^2 and relate it to $\Sigma_{ap}(t)$ using Equation 4.22. To estimate μ^2, we may use a stationary expression of the type discussed in Section 2.1,

$$\mu^2 = -\frac{\int_0^a dx\, \phi \dfrac{d^2\phi}{dx^2}}{\int_0^a dx\, \phi^2} + \lambda \frac{\int_0^a dx\, \phi_0 \phi^2}{\int_0^a dx\, \phi^2} \qquad (4.37)$$

A first estimate of μ^2 is obtained by using $\phi = \phi_0(x)$, yielding

$$
\mu^2 = - \frac{\displaystyle\int_0^a dx\, \phi_0 \frac{d^2\phi_0}{dx^2}}{\displaystyle\int_0^a dx\, \phi_0^2} + \lambda \frac{\displaystyle\int_0^a dx\, \phi_0^3}{\displaystyle\int_0^a dx\, \phi_0^2} \tag{4.38}
$$

The first term on the right is just the unperturbed eigenvalue B^2, so

$$
\mu^2 = B^2 + \lambda \frac{\displaystyle\int_0^a dx\, \phi_0^3}{\displaystyle\int_0^a dx\, \phi_0^2} \tag{4.39}
$$

We have thus succeeded in generating a trial function for the flux and estimates for the various parameters that are to be determined. If additional trial functions are desired, the procedure may be extended to yield second-order correction terms. (This was not done in the analysis of Section 4.6.) Expressions 4.31 and 4.19 can be used on the right side of the depletion Equation 4.2, which can then be integrated to yield a second-order expression for the fuel distribution. This fuel distribution may be used to obtain a second correction to the flux, and so forth. The procedure may be continued as long as is deemed necessary by the analyst.

4.6 A Comparison of Variational and Standard Calculations

In this section results are presented for the depletion problem of Section 4.1, obtained using the methods of Sections 4.2, 4.3, and 4.5. The standard results were obtained using the FEVER[28] depletion code for slab geometry with a single fuel. The numbers used were (with 1 barn = 10^{-24} cm^2)

$$D = 1 \text{ cm}$$

$$\sigma_a = 200 \text{ barns}$$

$$\nu\sigma_f = 400 \text{ barns}$$

$$B_0 = 2 \times 10^{19} \text{ cm}^{-3}$$

$$A_0 = 6.284 \times 10^{13} \text{ cm}^{-2} \text{ sec}^{-1}$$

$$a = 100 \text{ cm}$$

$$P = P'/(A_0 B_0) = 63.65 \text{ cm}$$
$$\text{(equivalent to 1280 watts/cm}^2)$$

The initial poison level was

$$\Sigma_{ap}(0) = 3.7605 \times 10^{-3} \text{ cm}^{-1}$$

The time period T under consideration has been chosen arbitrarily at 300 days. From the simple calculation

$$\frac{\delta_N}{B_0 N_0} \approx 1 - \sigma_a A_0 t \phi_0(x) \tag{4.40}$$

it can be seen that about 30 percent of the fuel at the center (x = 0)
will be depleted. In using FEVER, time steps of 50 days were
taken, so that the center fuel consumption was about 5 percent per
time step. The changes in fuel and flux distribution are shown in
Figures 4.1 and 4.2.

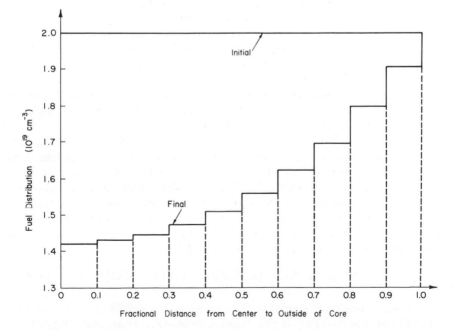

Figure 4.1. Initial and final fuel distributons from FEVER code.

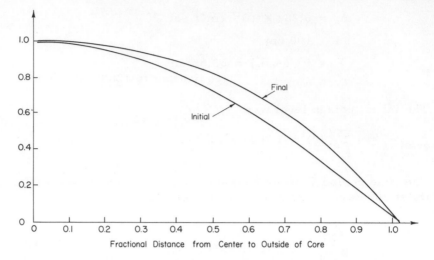

Figure 4.2. Initial and final flux distributions from FEVER code normalized to one at the center.

Before proceeding with the least-squares variational analysis, we must decide what value to use for the weighting constant k^2 in the functional Equation 4.14. In the discussion of the problem of relative weight in Section 3.7, it was pointed out that the need for a weighting factor tends to diminish as the ratio of system trial parameters increases. In the problems investigated here, this ratio is four to one. A small (relative to the lower limit of $k^2 = 1$) value of k^2 should be adequate. In addition, the results should not be overly sensitive to the exact value of k^2, provided the order of magnitude is appropriately chosen.

A variational calculation is made using the functional Equation 4.14 with $k^2 = 2$, using the FEVER trial function $\phi_F(x)$ as defined in Section 4.4. The poison level as a function of time is compared with the standard solution in Figure 4.3. The two results agree to within a fraction of 1 per cent. The fuel and flux distributions at $t = 100$, 200, and 300 days obtained by the two procedures are given in Figures 4.4 and 4.5. The standard and variational results are seen to be in good agreement in all cases.

The discrepancies between variational and standard flux solutions can be divided into shape and magnitude differences. Consider the variational flux at the end of life

$$\phi(x, T) = A_0[0.0033\phi_0(x) + 1.1338\phi_F(x)] \qquad (4.41)$$

The coefficient of the initial flux $\phi_0(x)$ is less than 0.3 of 1 per cent of the coefficient of the final flux $\phi_F(x)$, thus showing that the variational procedure yields accurate flux shapes.

It was stated above that a small weighting k^2 is advisable but

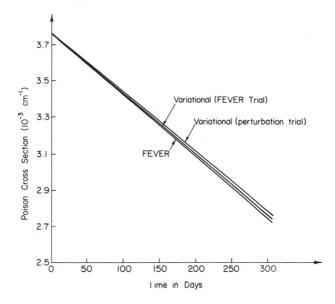

Figure 4.3. Variation of poison level with time by FEVER and by variational calculations using FEVER- and perturbation-generated trial functions.

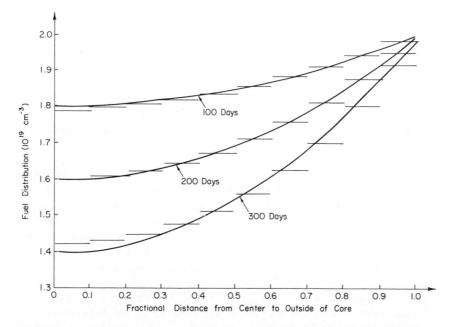

Figure 4.4. Fuel distribution at 100, 200, 300 days by FEVER (piecewise curves) and by variational (continuous curves) calculations using FEVER-generated trial functions.

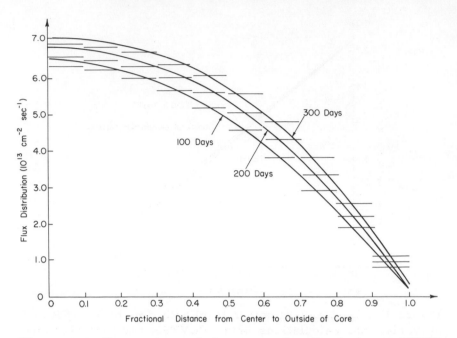

Figure 4.5. Comparison of average region fluxes from FEVER and variational flux distributions at 100, 200, 300 days using FEVER-generated trial functions.

that the results should not be too sensitive to the exact value of k^2. To check the validity of these statements, variational calculations are repeated for $k^2 = 1$, 3, and 4. The coefficients T_{11}, T_{21}, θ_{11}, θ_{21}, and ρ_1 (notation of Section 4.3) are given in Table 4.1 for the 4 values of k^2 investigated. The results are seen to be similar, even in the case of unit weighting ($k^2 = 1$). The results of the unit-weighted case, though, are seen to differ more from the results of the other 3 cases than the results of the other 3 cases differ from one another.

We also wish to make a variational calculation using the perturbation techniques of Section 4.5. Before making this calculation, however, let us see whether the Expression 4.31 is capable of describing the end-of-life flux $\phi_F(x)$. Equation 4.31 is evaluated for several values of λ. The function $\phi_P(x)$, obtained when the value of λ is three times the value given in Equation 4.23 is found to be very close to $\phi_F(x)$, as can be seen from Table 4.2. This function is used as the trial function $\psi_2(x)$.

Note that we have selected a value of λ, whereas earlier we had said that λ was the variable parameter to be determined. Actually, all we have done is to redefine the unknown parameter. Since

$$\phi(x) = A_0(a_1\phi_0 + a_2\phi_P) \tag{4.42}$$

Table 4.1. Variational Parameters for the Depletion Problem
for Several Weighting Constants k^2 (Relative Units)

k^2	T_{11}	T_{21}	θ_{11}	θ_{21}	ρ_1	$\Sigma_{ap}(T)$
			FEVER Trial			
1	-0.4000	0.4184	-0.05739	-0.05429	-0.1060	2.73
2	-0.3845	0.4374	-0.05783	-0.05816	-0.1016	2.77
3	-0.3846	0.4382	-0.05764	-0.05858	-0.0999	2.79
4	-0.3864	0.4352	-0.05776	-0.05780	-0.1077	2.71
			Perturbation Trial			
2	-0.3841	0.4412	-0.05784	-0.05854	-0.0992	2.79

Table 4.2. Comparison of Perturbation Trial ϕ_P and FEVER
Trial ϕ_F (Normalized to One at the Center)

x/a	ϕ_P	ϕ_F
0	1.0	1.0
0.1	0.9943	0.9940
0.2	0.9761	0.9741
0.3	0.9424	0.9387
0.4	0.8889	0.8850
0.5	0.8107	0.8100
0.6	0.7039	0.7090
0.7	0.5663	0.5806
0.8	0.5988	0.4237
0.9	0.2067	0.2411
1.0	1.01×10^{-6}	0.0409

Table 4.3. Comparison of Fuel and Flux Distributions Obtained with FEVER and Perturbation-Generation Trial Functions ϕ_F and ϕ_P

x/a	Fuel ($\times 10^{-19}$)		Flux ($\times 10^{-13}$)	
	With ϕ_P	With ϕ_F	With ϕ_P	With ϕ_F
At 100 Days				
0	1.800	1.800	6.595	6.572
0.1	1.800	1.801	6.536	6.512
0.2	1.806	1.807	6.347	6.321
0.3	1.815	1.816	6.027	5.996
0.4	1.828	1.829	5.568	5.538
0.5	1.846	1.847	4.968	4.947
0.6	1.868	1.868	4.227	4.224
0.7	1.895	1.894	3.352	3.373
0.8	1.926	1.924	2.357	2.498
0.9	1.961	1.957	1.271	1.347
1.0	1.997	1.992	0.130	0.227
At 200 Days				
0	1.600	1.599	6.905	6.859
0.1	1.601	1.603	6.855	6.808
0.2	1.612	1.613	6.695	6.641
0.3	1.630	1.632	6.412	6.352
0.4	1.656	1.658	6.000	5.931
0.5	1.692	1.693	5.408	5.365
0.6	1.737	1.737	4.652	4.644
0.7	1.790	1.788	3.717	3.760
0.8	1.852	1.848	2.617	2.717
0.9	1.921	1.915	1.381	1.535
1.0	1.994	1.986	0.065	0.260
At 300 Days				
0	1.397	1.400	7.215	7.146
0.1	1.402	1.404	7.174	7.104
0.2	1.418	1.420	7.042	6.961
0.3	1.445	1.448	6.798	6.708
0.4	1.485	1.488	6.411	6.323
0.5	1.538	1.540	5.847	6.784
0.6	1.605	1.605	5.076	5.065
0.7	1.686	1.682	4.083	4.147
0.8	1.779	1.772	2.876	3.026
0.9	1.882	1.872	1.491	1.722
1.0	1.991	1.978	0.008	0.292

$$\phi_P = \phi_0 - \frac{\lambda a^2}{2\pi^2} \phi_1 \tag{4.43}$$

with $\phi_1(x)$ given by Equation 4.36, we may either specify λ, use ϕ_P as the trial function and then determine a_1 and a_2 or else we may let

$$\phi(x) = A_0(b_1\phi_0 + b_2\phi_1) \tag{4.44}$$

and determine b_1 and b_2. The parameters are related by

$$b_1 = a_1 + a_2 \tag{4.45}$$

$$b_2 = -\frac{\lambda a^2}{2\pi^2} a_2 \tag{4.46}$$

We have chosen the former course, although the latter course must yield the same results.

A variational calculation is made with $k^2 = 2$. The parameters T_{11}, T_{21}, θ_{11}, θ_{21}, and ρ_1 and the end-of-life poison level $\Sigma_{ap}(T)$ are given in Table 4.1. The poison level as a function of time is plotted in Figure 4.3, where it is seen to be close to the poison levels yielded by the FEVER trial and by the standard procedure. The fuel and flux distributions obtained using the perturbation trial $\phi_P(x)$ are too close to those obtained using $\phi_F(x)$ to permit plotting in Figures 4.4 and 4.5. The results at 100, 200, and 300 days are compared for the trial functions $\phi_P(x)$ and $\phi_F(x)$ in Table 4.3.

One of the most important parameters of interest in a depletion study is the core lifetime. Since the results for the poison level are accurate, it is to be expected that the value of the lifetime would also be accurate. Some difficulty may arise in cases where the poison level varies slowly with time, as can be the case when additional fuel is produced during core life (as in a 'converter' reactor). A small error in poison level may then result in a large error in lifetime. Suppose, for example, that we consider end of life to be when the poison level reaches the value 2.91. This corresponds to a lifetime of 250 days. With $k^2 = 2$, the FEVER trial-function and perturbation trial-function methods yield lifetimes of 258 and 263 days, respectively. With $k^2 = 1$, 3, and 4, the FEVER trial function yields lifetimes of 247, 260, and 244 days, respectively, thus illustrating the effect of the choice of k^2 on the core lifetime.

Chapter 5

DISCUSSION AND SUGGESTIONS FOR FURTHER STUDY

5.1 General Variational Problems

In Section 3.1, a set of criteria was set up by which to choose the "best" functional for use in generalized variational procedures. It was pointed out in Section 3.8, however, that in specific problems, certain specific criteria might arise that add to or supersede the criteria of Section 3.1. We might then consider using other functionals in a generalized variational procedure.

Suppose, for example, that we are interested in finding a stationary expression (not involving adjoint functions) for the eigenvalue λ of the non-self-adjoint equation

$$L\phi - \lambda M\phi = 0 \tag{5.1}$$

Let us operate on Equation 5.1 with the adjoint operator L^*, yielding

$$L^*L\phi - \lambda L^*M\phi = 0 \tag{5.2}$$

If the operator L^*M is self-adjoint, then Equation 5.2 is self-adjoint, and a stationary expression, whose Euler equation is Equation 5.2, is

$$\lambda = \frac{\displaystyle\int_a^b dx\ \phi L^*L\phi}{\displaystyle\int_a^b dx\ \phi L^*M\phi} \tag{5.3}$$

It is suggested that when we are dealing with specific problems we consider not only the specific variational methods discussed in this work but also the possibility of using the general approach of Section 3.1 to generate variational principles suited to the particular problems at hand.

A second topic for study is the use of semidirect methods (as discussed in Section 2.2) with generalized variational methods. (Semidirect methods were not used in connection with least-

squares procedures in this work.)   One complication is that since
the equations to be solved are differential equations, and not algebraic equations, "extraneous" boundary conditions must be applied.

5.2 Depletion and Related Problems

From the mathematical point of view, the depletion problem
analyzed in Chapter 4 is quite complicated, since it involves a
nonlinearity, a constraint, and several dependent and independent
variables.   Thus, the results of that chapter clearly indicate the
applicability of least-squares variational methods to complicated
problems.

From the physical point of view, however, the depletion model
discussed in Section 4.1 is greatly simplified.   Such important
matters as the buildup of fission products during core life, the
conversion of fertile material to fissile material, the change in
energy spectrum during core life (i. e., the need for more than
one energy group), the presence of a 'reflector' about the core,
and the possibility of a multiregion core with more complicated
control arrangements were not considered.

As was pointed out in Section 4.1, the forms of the equations
are not changed by these refinements: more energy groups implying more equations such as Equation 4.1, fission products and
additional fuel isotopes implying more equations such as Equation 4.2 and additional terms in the constraint Equation 4.3.   Although the forms of the equations are not changed, the forms of
the solutions may be changed considerably.   In the model of Section 4.1, for example, the monotonic fuel depletion results in a
monotonic (and practically constant) decrease in the poison level.
In many nuclear reactors, however, plutonium production at the
beginning of life more than offsets the uranium-235 consumption
causing an initial increase in reactivity.   The poison level should,
therefore, increase to a maximum value before beginning to decrease.   Considerations such as this indicate that more care must
be taken in the selection of trial functions when analyzing more
complicated depletion problems, particularly for trial functions
in the time domain.

When dealing with multiregion problems, we are faced with the
necessity for satisfying interface boundary conditions for the flux.
These are continuity of flux and current, the current being given
by Fick's law

$$\underset{\sim}{J} = -D\underset{\sim}{\nabla}\phi \tag{5.4}$$

Equation 5.4 will not appear as a natural boundary condition from
a least-squares principle such as Equation 4.14.   It is possible,
however, to add to the functional terms that will make the bound-

ary conditions be satisfied in the least-squares sense.

Consider, for example, a two-region, one-dimensional problem with the fluxes in the two regions denoted by $\phi_1(x, t)$ and $\phi_2(x, t)$. At the interface x_I, these fluxes must satisfy

$$\phi_1(x_I, t) = \phi_2(x_I, t) \tag{5.5}$$

$$D_1 \left. \frac{\partial \phi_1}{\partial x}(x, t) \right|_{x=x_I} = \left. D_2 \frac{\partial \phi_2}{\partial x} \right|_{x=x_I} \tag{5.6}$$

In each region, the flux can be expanded in trial functions

$$\phi_1(x, t) \approx \sum_{i=1}^{N} a_{1i} \psi_i(x, t) \tag{5.7}$$

$$\phi_2(x, t) \approx \sum_{i=1}^{N} a_{2i} \eta_i(x, t) \tag{5.8}$$

Continuity of flux and current can be expressed in the mean-square sense by adding the terms

$$\int_0^T dt \, [\phi_1(x_I, t) - \phi_2(x_I, t)]^2 \tag{5.9}$$

$$\int_0^T dt \left[D_1 \frac{\partial \phi_1}{\partial x}(x_I, t) - D_2 \frac{\partial \phi_2}{\partial x}(x_I, t) \right]^2 \tag{5.10}$$

to the functional. In multidimensional problems, these terms would involve surface integrals; that is, Equations 5.9 and 5.10 would become

$$\int_0^T dt \int dS \, [\phi_1(S, t) - \phi_2(S, t)]^2 \tag{5.11}$$

$$\int_0^T dt \int dS \, [D_1 \underset{\sim}{\nabla} \phi_1(S, t) - D_2 \underset{\sim}{\nabla} \phi_2(S, t)]^2 \tag{5.12}$$

This procedure may be looked upon as a generalization of the addition of boundary terms to a functional to achieve certain boundary conditions. This squared boundary-condition approach though, has an additional potential application. It may be used when, as a result of a slightly irregular boundary surface or for some other reason, it is difficult to generate a trial function that satisfies the specified boundary conditions exactly.

A reflector might be treated as an additional region, and the multiregion technique just described could be used. A more simplified approach, useful in large systems, would be to consider an unreflected core which has a reflector savings.[30]

More complicated control systems should not offer too much difficulty. As an illustration, consider a reactor such as that of Section 4.1, except that instead of being controlled by a variable uniform poison, it is controlled by an extractable constant poison. This model corresponds to a moving bank of control rods. Initially, there is a uniform poison in the core with cross section Σ_{ap}. As power is produced, the poison is withdrawn to maintain criticality at the design-power level, and we obtain a two-region problem with a variable interface $x_I(t)$ such that the poison cross section is

$$\left. \begin{array}{ll} \Sigma_{ap} & \text{for} \quad 0 \le x \le x_I(t) \\ \\ \\ 0 & \text{for} \quad x_I(t) \le x \le a \end{array} \right\} \qquad (5.13)$$

The boundary terms Equations 5.9 and 5.10 are to be applied. The control variable in this case is the interface position $x_I(t)$. The functional is therefore to be minimized with respect to variations in $x_I(t)$.

An increase in the complexity of the problem causes an increase in the number of trial parameters to be determined. Nel[21] reports that use of an advanced gradient method enabled him to solve for each of 48 unknowns in a nonlinear system of equations to within 1 percent of its exact value, using about 2 1/2 minutes on the IBM-7094 computer; so solving the algebraic equations should not be a limiting factor.

Another interesting attribute of the variational method is its ability to provide additional information with only a small amount of additional effort. The most time-consuming part of the analysis, particularly in 2 and 3 spatial dimensions, is usually the evaluation of the various integrals. Suppose we wished to determine the effects of small changes in some of the system parameters on the results. If the same trial functions are deemed adequate for all cases, then the integrals need be evaluated for only one case and then saved for use in analyzing the other cases. Whereas the time consumed for the first case is determined by the time consumed in evaluating the integrals, the time consumed in each additional case is dependent on the time needed to solve the nonlinear equations. Even this latter time will be generally smaller than for the first case, because the results for the first case should provide good initial estimates for the parameters in the other cases.

From the successful results of the Section 5.1, and from this discussion, it appears worthwhile to attempt to use the least-squares variational method for a "full-fledged" depletion problem. Further work on the development of perturbation techniques for generating trial functions should also be useful.

Because of the similarities in the forms of the equations for the three problems of Section 4.1, it should prove profitable to apply least-squares variational procedures to the xenon oscillation and short-time kinetics problems. There is, however, a distinctive feature about these problems that is not present in the depletion problem. This feature is the presence of transient effects.

The depletion process is a slow one, the flux distribution existing in a quasi-steady state. The process occurs naturally, not as the result of an externally induced change in the system properties. In the xenon and kinetics problems, however, we investigate the response of the system to perturbations. Particularly in the kinetics problem, the response has two phases. One phase is the initial transient reaction to the perturbation, in which large changes in flux magnitude and severe distortion in flux distribution can take place. Trial functions must be capable of describing the severe distortions. The second phase, the approach to the "persisting" solution, should provide less difficulty.

Appendix A

THE SIGNIFICANCE OF THE ADJOINT FUNCTION

An understanding of the significance that may be attributed to
the adjoint function can be very useful in the generation of adjoint trial functions, as is illustrated in Section 2.5. Such an
understanding is also helpful for other applications, as can be
seen from the discussion of Section 2.1.

Much work[14,15,26] has been done in recent years, particularly
with regard to linear problems of nuclear reactor physics, in
attempting to understand the adjoint functions. Relatively little
work has been done, however, outside the context of reactor
physics or in connection with nonlinear problems.

This appendix has two objectives: one pertaining to linear problems, the other pertaining to nonlinear problems. With regard
to linear problems, the objective is to develop in Section A.1 the
significance of the adjoint function for an arbitrary system. It
is felt that a more general viewpoint is thereby obtained toward
the results. To illustrate possible application to problems outside of reactor physics, it is shown that the adjoint temperature
in the heat-conduction problem may be related to the maximum
thermal stress in the system. The approaches taken in the analysis of linear systems are based on treatments of linear reactor
physics problems given by Selengut[26] and Lewins.[15] In nonlinear
problems, the object is to extend in Section A.2 the results of the
Hamilton-Jacobi theory,[7] where the adjoint function (or "conju-
gate" variable) is found to be the effect on the functional of a small
change in boundary conditions in one-dimensional problems,[5] to
positions off the boundaries, and to multidimensional problems.

A.1 Linear Problems and Green's Functions

A.1.1 Inhomogeneous Problems. An inhomogeneous equation
is one which has either a source term or nonzero boundary conditions. Since nonzero boundary condtions can be treated as
singular sources, a general inhomogeneous equation is

$$H\phi - f = 0 \qquad (A.1)$$

with homogeneous boundary conditions. The dual functional

$$J = \int_a^b dx \, [\phi^*(H\phi - f) - g\phi] \qquad (A.2)$$

yields, upon variation of ϕ^* and ϕ, Equations A.1 and

$$H^*\phi^* - g = 0 \qquad (A.3)$$

the adjoint operator H^* being defined in Section 2.3. In this appendix we shall refer to Equation A.1 and ϕ as the system equation and the system function, respectively, and shall refer to Equation A.3 and ϕ^* as the adjoint equation and the adjoint function, respectively.

It should be noted that Equation A.2 can be rewritten in the form

$$J = \int_a^b dx \, [\phi(H^*\phi^* - g) - f\phi^*] \qquad (A.4)$$

When Equations A.1 and A.3 are satisfied, the value of the functional J is given by each of the two expressions

$$J = -\int_a^b dx \, g(x) \, \phi(x) \qquad (A.5)$$

$$J = -\int_a^b dx \, f(x) \, \phi^*(x) \qquad (A.6)$$

Since the choice of the inhomogeneous term $g(x)$ in the adjoint equation is, to a very large extent, arbitrary, we have a great deal of freedom in choosing a quantity of interest J, related to the system, which is to be obtained with second-order accuracy.

To see what significance is possessed by the adjoint function, let us consider the effect of a unit source

$$f(x) = \delta(x - x_0) \qquad (A.7)$$

Substitution in Equation A.6 yields

$$J = -\phi^*(x_0) \qquad (A.8)$$

Comparison of Equation A.8 with Equation A.5 indicates that the adjoint function is the contribution of a unit source to the quantity of interest. In this sense, the adjoint function is a Green's function[19] for some quantity of interest related to the system, the quantity being determined by the choice of the adjoint in-homogeneity.

Because of this type of interpretation, the adjoint function has been referred to as an "importance" function.[14] We could say that regions in which the adjoint function is relatively large are more important than regions in which the adjoint function is rel-atively small, because a unit source has a greater effect in the large-adjoint regions.

When Equation A.1 is an initial-value problem, then the ad-joint Equation A.3 must be a final-value problem if the bilinear concomitant (see Section 2.3) is to vanish at the initial and final times. Somewhat different interpretations are obtained depending on whether the adjoint inhomogeneity is taken to be a source or a final-value condition.

Consider, for example, the equation

$$H(x, t)\phi - f(x, t) = 0 \qquad\qquad (A.9)$$

and the functional

$$J = \int_0^T dt \int_a^b dx \, [\phi^*(H\phi - f) - g(x, t) \, \phi(x, t)] \qquad (A.10)$$

The value of the functional is each of the following:

$$J = -\int_0^T dt \int_a^b dx \, g(x, t) \, \phi(x, t) \qquad\qquad (A.11)$$

$$J = -\int_0^T dt \int_a^b dx \, f(x, t) \, \phi^*(x, t) \qquad\qquad (A.12)$$

Let f be the unit source

$$f(x, t) = \delta(t - t_0) \, \delta(x - x_0) \qquad\qquad (A.13)$$

Equation A.12 thus becomes

$$J = -\phi^*(x_0, t_0) \qquad\qquad (A.14)$$

In a physical system, a source can affect the future, but not the past. With the source Equation A.13, $\phi(x, t)$ must be zero for times preceding t_0, and Equation A.11 becomes

$$J = -\int_{t_0}^{T} dt \int_{a}^{b} dx\ g(x, t)\ \phi(x, t) \tag{A.15}$$

The adjoint function is therefore the integrated contribution to the quantity of interest for the time interval $t_0 < t \leq T$. If the adjoint homogeneity is a final-value condition, then

$$g(x, t) = \phi_T^*(x)\ \delta(t - T) \tag{A.16}$$

and Equation A.15 becomes

$$J = -\int_{a}^{b} dx\ \phi_T^*(x)\ \phi(x, T) \tag{A.17}$$

and the adjoint function is the contribution to some quantity at the final time T.

In the area of nuclear reactor physics, Selengut[26] has pointed out that if $g(x)$ in Equation A.3 is a cross section (reaction-rate probability), then the adjoint neutron flux is the reaction rate in the system because of a unit neutron source. Lewins[15] has pointed out that if

$$\phi_T^*(x) = 1 \tag{A.18}$$

then the adjoint neutron density can be related to the response of a set of detectors.

Many other applications are possible because of the flexibility permitted in choosing the adjoint inhomogeneity. Suppose, for example, that we are interested in the heat generated at a certain point x_0 (a "hot spot," for example) in a reactor. The heating rate is proportional to the fission cross section Σ_f and to the flux ϕ. We would therefore choose

$$g(x, t) = \Sigma_f(x)\ \delta(x - x_0) \tag{A.19}$$

Note that f also may be either a source or initial value. In the latter case we would have

$$f(x) = \phi_0(x)\ \delta(t) \tag{A.20}$$

We would then speak of the adjoint function as the effect of a unit initial condition on the quantity of interest.

The Green's function interpretation permits two interesting applications of the adjoint function. Suppose that we were interested in making a set of survey calculations, to calculate a "figure of merit" of the system for each of several sources f(x)

$$J = \int_a^b dx\; g(x)\; \phi(x) \tag{A.21}$$

We would seek to obtain a Green's function formulation in terms of the source f(x)

$$J = \int_a^b dx\; G(x)\; f(x) \tag{A.22}$$

where G(x) is the Green's function. We know, however, that G is simply the adjoint function $\phi^*(x)$ if the adjoint inhomogeneity is g(x).

The second application is to apply the insight given by the adjoint function to the selection of the external source f, that is, to use the adjoint function as a design variable. Since the adjoint function tells us where a unit source is most effective in contributing to the figure of merit, we can design the source f(x) so as to maximize this figure of merit.

In a sense, then, the adjoint function ϕ^* might be considered as a more fundamental description of the system than is the system function ϕ. The system function tells us how the system will behave for a given external source. The adjoint function, on the other hand, tells us how the system responds to sources in general. This is especially so when (as will usually be the case if the quantity of interest J is a property of the system) the adjoint source g(x) is composed only of intensive and extensive system properties.

A.1.2 Application to Heat Conduction and Thermal Stress. In Section A.1.1 the significance of the adjoint function was discussed for an arbitrary system, and some applications to reactor physics were noted. Here we present an example outside of reactor physics (although pertinent to reactor engineering).

Let us consider the conduction of heat in a long cylinder of radius R with heat source H(r). The temperature difference T(r) between the radii r and R is given by an equation of the form

$$LT(r) = H(r) \tag{A.23}$$

We might be interested in deciding how to distribute the heat source, that is, to choose the form of $H(r)$, in order that the system conform to thermal stress-design limitations. The radial, tangential, and axial stresses are given respectively by[2]

$$\sigma_{rr}(r) = \frac{\alpha E}{1 - \nu} \left[\frac{1}{R^2} \int_0^R r\,T(r)\,dr - \frac{1}{r^2} \int_0^r r_1 T(r_1)\,dr_1 \right]$$

(A.24)

$$\sigma_{\theta\theta}(r) = \frac{\alpha E}{1 - \nu} \left[\frac{1}{R^2} \int_0^R r\,T(r)\,dr + \frac{1}{r^2} \int_0^r r_1 T(r_1)\,dr_1 - T(r) \right]$$

(A.25)

$$\sigma_{zz}(r) = \frac{\alpha E}{1 - \nu} \left[\frac{2}{R^2} \int_0^R r\,T(r)\,dr - T(r) \right]$$ (A.26)

where α, E, and ν are, respectively, the thermal expansion coefficient, the elastic constant, and Poisson's ratio. Noting that

$$\lim_{r \to 0} \frac{1}{r^2} \int_0^r r_1 T(r_1)\,dr_1 = \frac{1}{2} T(0)$$ (A.27)

and that the maximum radial and axial stresses must occur at the center of the cylinder, we might be interested in the adjoint equation

$$L^* T^* = \frac{\alpha E}{1 - \nu} \frac{1}{R^2} - \frac{1}{2} \delta(r)$$ (A.28)

because the functional

$$-J = \frac{\alpha E}{1 - \nu} \left[\frac{1}{R^2} \int_0^R dr\,r\,T(r) - \frac{1}{2} T(0) \right]$$ (A.29)

is the maximum radial stress and half of the maximum axial stress. We would solve Equation A.28 for $T^*(r)$ and then evaluate

$$J = - \int_0^R dr\,H(r)\,T^*(r)$$ (A.30)

for the maximum stresses for each of several heat sources H. The adjoint temperature $T^*(r)$ is the Green's function for the maximum thermal stress.

Knowledge of the adjoint temperature can be helpful in modifying heat sources so as to improve their performance with respect to thermal stresses. We would attempt to reduce the heat source where the adjoint temperature is relatively large and to increase it where the adjoint temperature is relatively small.

A.1.3 <u>Homogeneous Problems.</u> Let us now consider homogeneous equations with homogeneous boundary conditions. (Homogeneous equations with inhomogeneous boundary conditions can be transformed into inhomogeneous equations with homogeneous boundary conditions and are therefore covered by the discussion of Section A.1.1.)

Homogeneous steady-state problems have received considerable attention with respect to the significance of the adjoint function. This state of affairs is probably because of the widespread use of adjoint functions in perturbation theory.[14] It was in connection with this type of problem that the "importance" concept was originally introduced.

A homogeneous problem may be expressed in the form

$$\phi(x) \ = \ \int_a^b \ G(x, x') \ \phi(x') \ dx' \tag{A.31}$$

and its adjoint problem may be written noting that the kernel of the adjoint equation is the transpose of the kernel of the system equation,[20]

$$\phi^*(x) \ = \ \int_a^b \ G(x', x) \ \phi^*(x') \ dx' \tag{A.32}$$

In Section A.1.1 it was pointed out that the adjoint function was the contribution to a quantity of interest by a unit source. In homogeneous problems, however, we do not have the flexibility in choosing the source that we have in inhomogeneous problems.

The amount of the quantity ϕ appearing at x' because of a unit amount of ϕ at x is $G(x', x)$. Let us define $\phi^*(x')$ as the contribution to some quantity of interest caused by a unit amount at x' and see if the definition is consistent with the adjoint equation. The contribution caused by the amount of ϕ appearing at x' because of the unit amount at x is $G(x', x) \ \phi^*(x')$. The total contribution in the system because of a unit amount at x is, therefore,

$$\int_a^b \phi^*(x') \, G(x', x) \, dx' \tag{A.33}$$

This, however, is again the definition of $\phi^*(x)$, so

$$\phi^*(x) = \int_a^b dx' \, G(x', x) \, \phi^*(x') \tag{A.34}$$

which is the adjoint equation. The definition is thus seen to be consistent with the adjoint equation. In the Green's function interpretation, therefore, the source is a unit amount of ϕ and the "quantity of interest" is arbitrary. This conclusion is consistent with Selengut's observation[27] that "the neutron importance with respect to any physical process that can occur in a (nuclear) reactor will be the same function except for a scale factor."

There is another point regarding homogeneous problems that should be noted. Homogeneous problems are characteristic value-characteristic function problems. As a result, we cannot really speak in an arbitrary manner about the response resulting from a unit amount of ϕ. The distribution $\phi(x)$ must be proportional to a characteristic function (eigenfunction). Consequently, when we speak about the effect because of a unit amount of ϕ, it is implicit that this unit amount of ϕ is a constituent of a distribution proportional to an eigenfunction. It is because of this that the response function, that is, the adjoint function, is itself an eigenfunction (of the adjoint operator) characterized by the same characteristic value (eigenvalue).

A.2 Nonlinear Problems

A.2.1 Relevant Differences Between Linear and Nonlinear Problems. In linear problems, it was pointed out that the adjoint function could be interpreted as a Green's function, that is, the contribution of a unit source to a quantity of interest. In nonlinear problems, however, effects cannot be superposed (the contribution of a unit source is affected by the presence of other sources), and the Green's function interpretation cannot be applied.

The relationship between the adjoint and system equations is significantly changed by the presence of nonlinearities. In nonlinear problems, the adjoint operator H^* is a function of ϕ, but the adjoint equation itself is linear in ϕ^*. (This is the case because in the functional

$$J = \int_a^b dx \, [\phi^*(H\phi - f) - g\phi] \tag{A.35}$$

ϕ^* appears linearly, but, since H is a nonlinear operator, ϕ appears nonlinearly.) Thus, the adjoint equation cannot be solved unless the system function ϕ is known. In nonlinear problems, therefore, the adjoint equation cannot serve as an alternative description of the system, as is the case in linear problems. It also follows that a nonlinear operator cannot be self-adjoint.

A.2.2 An Importance Interpretation in One-dimensional Problems. In one-dimensional problems, the Hamilton-Jacobi theory[7] can be applied to yield an importance interpretation. Since an n-th—order ordinary differential equation can be expressed as a set of n simultaneous first-order equations,[23] a general one-dimensional problem may be written in the vector notation

$$\dot{\underset{\sim}{y}} = \underset{\sim}{f}(x, \underset{\sim}{y}) \tag{A.36}$$

where $\underset{\sim}{y}$ and $\underset{\sim}{f}$ are n-vectors and the dot denotes differentiation with respect to x. A dual functional yielding Equation A.36 is

$$J = \int_a^b dx \, [L(x, \underset{\sim}{y}) + \langle \underset{\sim}{p}, (\underset{\sim}{f} - \dot{\underset{\sim}{y}}) \rangle] \tag{A.37}$$

where $\underset{\sim}{p}$ is the adjoint vector and $\langle , \rangle$ denotes scalar product. The value of the functional is

$$\int_a^b dx \, L(x, \underset{\sim}{y}) \tag{A.38}$$

The significance of Equation A.38 is determined by the analyst when he chooses $L(x, \underset{\sim}{y})$. The adjoint equation is

$$\dot{\underset{\sim}{p}} = -\frac{\partial}{\partial \underset{\sim}{y}} \langle \underset{\sim}{p}, \underset{\sim}{f} \rangle - \frac{\partial}{\partial \underset{\sim}{y}} L(x, \underset{\sim}{y}) \tag{A.39}$$

where $\partial/\partial \underset{\sim}{y}$ is the vector operator

$$\begin{array}{c} \dfrac{\partial}{\partial y_1} \\ \cdot \\ \cdot \\ \cdot \\ \dfrac{\partial}{\partial y_n} \end{array} \tag{A.40}$$

In Hamilton-Jacobi theory we define a "Hamiltonian" of the form

$$\mathcal{H}(x, \underset{\sim}{y}, \underset{\sim}{p}) \;=\; L(x, \underset{\sim}{y}) + \langle \underset{\sim}{p}, \underset{\sim}{f} \rangle \tag{A.41}$$

in which case the Equations A.36 and A.39 become

$$\underset{\sim}{\dot{y}} \;=\; \frac{\partial \mathcal{H}}{\partial \underset{\sim}{p}} \tag{A.42}$$

$$\underset{\sim}{\dot{p}} \;=\; - \frac{\partial \mathcal{H}}{\partial \underset{\sim}{y}} \tag{A.43}$$

Let the problem of making the functional, Equation A.37, stationary be imbedded in the general class of problems with the functional

$$J \;=\; \int_{z}^{b} dx \left[\mathcal{H}(x, \underset{\sim}{y}, \underset{\sim}{p}) - \langle \underset{\sim}{p}, \underset{\sim}{\dot{y}} \rangle \right] \tag{A.44}$$

where z is a variable lower limit. Equation A.37 is a special case of Equation A.44 with $z = a$. The functional J becomes a function of z and of the boundary values $y(z)$. It can be shown[1,5] that J satisfies the Hamilton-Jacobi equation

$$- \frac{\partial J}{\partial z} \;=\; L(z, \underset{\sim}{y}) + \langle \frac{\partial J}{\partial \underset{\sim}{y}}, \underset{\sim}{f} \rangle \tag{A.45}$$

and that the following relations are satisfied

$$\underset{\sim}{p}(z) \;=\; \frac{\partial J}{\partial \underset{\sim}{y}}(z) \tag{A.46}$$

$$\mathcal{H}(z, \underset{\sim}{y}, \underset{\sim}{p}) \;=\; - \frac{\partial J}{\partial z} \tag{A.47}$$

The adjoint function and the Hamiltonian thus describe the effects of changes in boundary conditions and in the size of interval (z, b) on the functional J. It is also clear from Equation A.45 that the functional J is a linear function (that is, the Hamilton-Jacobi equation is a linear partial differential equation) of the boundary values and of the interval size, thus revealing a linear facet of nonlinear problems. For small changes δy, we may write

$$\delta J \;=\; \langle \underset{\sim}{p}(z), \delta \underset{\sim}{y}(z) \rangle \tag{A.48}$$

since the linearity of the Hamilton-Jacobi equation indicates that we may superpose the effects of the changes in the various components of $\delta \underset{\sim}{y}$.

In investigating the interpretation of the adjoint function, however, we are interested in the significance of the adjoint function
in the whole domain of the problem, not just at the boundaries.
In extending the interpretation to positions off the boundaries, we
shall make use of the principle of optimality,[1] which states that
portions of optimal trajectories (that is, of functions which are
solutions of Euler equations) are themselves optimal trajectories.
If, for example, the function $y_0(x)$ makes the functional

$$J = \int_a^b dx\ F(x, y)$$ (A.49)

stationary, then, if $a < c < b$ and

$$J_1 = \int_a^c dx\ F(x, y)$$ (A.50)

$$J_2 = \int_c^b dx\ F(x, y)$$ (A.51)

with the boundary condition

$$y(c) = y_0(c)$$ (A.52)

then $y_0(x)$ also makes Equations A.50 and A.51 stationary and the
values of the functionals obey the relation

$$J = J_1 + J_2$$ (A.53)

Let us suppose that we wish to know the significance of $p(x)$ for
some x in the interval $a < x < b$. Since portions of optimal trajectories are themselves optimal trajectories, the problem may
be broken into two separate problems, one dealing with the interval (a, x), the other with the interval (x, b). Suppose that a small
change $\delta y(x)$ is introduced. The adjoint function $p(x)$ represents
the effect of this change on that part of the functional evaluated
over the interval (x, b). Since the functional evaluated over the
interval (a, b) is insensitive to changes in the function $y(x)$, the
effect on the interval (a, x) is equal and opposite to the effect on
the interval (x, b).

Let us illustrate the development of this section with an example. The problem of neutron diffusion under the effects of

xenon poisoning (discussed in Section 3.6.1) is of the form

$$\frac{d^2\phi}{dx^2} = h(\phi, x) \tag{A.54}$$

To put this equation into the "canonical form," Equation A.36, let

$$y_1 = \phi \tag{A.55}$$

$$y_2 = \frac{dy_1}{dx} \tag{A.56}$$

whereby Equation A.54 becomes the vector equation

$$\begin{bmatrix} \dot{y}_1 \\ \dot{y}_2 \end{bmatrix} = \begin{bmatrix} y_2 \\ h \end{bmatrix} \tag{A.57}$$

or

$$\dot{\underset{\sim}{y}} = \underset{\sim}{f}(x, \underset{\sim}{y}) \tag{A.58}$$

Consider the functional

$$J = \int_a^b dx \left[g(x)\, y_1 + p_1(y_2 - \dot{y}_1) + p_2(h - \dot{y}_2) \right] \tag{A.59}$$

If g is taken to be the cross section for some process (for example, fission) then the value of the functional is the reaction rate for that process. The adjoint equations are

$$\dot{p}_1 = -g - \frac{\partial h}{\partial y_1}\, p_2 \tag{A.60}$$

$$\dot{p}_2 = -p_1 \tag{A.61}$$

If a perturbation δy_1 is introduced at the position x, then the reaction rate will be changed by $p_1(x)\, \delta y_1$ in the interval (x, b) and an equal or opposite change will occur in the interval (a, x).

The adjoint function may be interpreted as an importance function in nonlinear problems, as well as in linear problems. A given perturbation has a relatively large effect on the quantity of interest in a region where the adjoint function is large, a relatively small effect where the adjoint function is small.

Let us conclude the discussion of the one-dimensional case by relating the vector adjoint equation to the scalar adjoint equation. Referring to Equations A.60 and A.61, let

$$\phi^* = p_2 \qquad (A.62)$$

This leads to

$$\frac{d^2\phi}{dx^2} + \frac{\partial h}{\partial y_1}\phi^* + g = 0 \qquad (A.63)$$

which is easily seen to be adjoint to Equation A.54. Note that the scalar adjoint function corresponds to the last component of the vector adjoint function (the scalar system function corresponds to the first component of the vector system function). This is because the taking of adjoints generally has a "transposing" effect.

A.2.3 Multidimensional Problems. In dealing with multidimensional problems, a procedure somewhat different from the Hamilton-Jacobi-theory approach (applicable only to one-dimensional problems) is needed. A clue as to how to proceed is provided by another method for obtaining the Relations A.46 and A.45. In dealing with the general variation of a functional (varying both the function and the interval) it can be shown[8] that when the Euler Equations A.36 and A.39 are satisfied, the variation with respect to the function $\underset{\sim}{y}$ becomes

$$\delta J = -\langle \underset{\sim}{p}, \delta \underset{\sim}{y}\rangle\Big|_a^b + (\mathscr{H}\delta x)\Big|_a^b \qquad (A.64)$$

If the upper limit conditions are satisfied, we again obtain

$$\underset{\sim}{p}(a) = \frac{\partial J}{\partial \underset{\sim}{y}(a)} \qquad (A.65)$$

$$\mathscr{H}(a) = -\frac{\partial J}{\partial a} \qquad (A.66)$$

which are the same as Equations A.46 and A.47. Whereas Equations A.46 and A.47 were obtained by considering variations along optimal trajectories, Equations A.65 and A.64 were obtained for the more general case of arbitrary variations. Since we have already seen how an interpretation of the adjoint function in terms of effects of boundary-condition perturbation could be extended to positions off the boundaries, the procedure leading to Equation A.64 may be applied to multidimensional problems to obtain a boundary interpretation, and this boundary interpretation extended to the "interior." For convenience, we shall consider the fixed-interval case.

Let us proceed by example. Suppose we have a quantity of interest of the form

$$J = \int_{a_1}^{b_1} \int_{a_2}^{b_2} L(\phi, x, y) \, dx \, dy \tag{A.67}$$

with ϕ given by the partial differential equation

$$\frac{\partial^2 \phi}{\partial x^2} + \frac{\partial^2 \phi}{\partial y^2} = f(x, y, \phi) \tag{A.68}$$

Let

$$\phi_1 = \phi \tag{A.69}$$

$$\phi_2 = \frac{\partial \phi}{\partial x} \tag{A.70}$$

$$\phi_3 = \frac{\partial \phi}{\partial y} \tag{A.71}$$

The partial differential Equation A.68 thus becomes the set of simultaneous first-order partial differential equations

$$\frac{\partial \phi_1}{\partial x} = \phi_2 \tag{A.72}$$

$$\frac{\partial \phi_1}{\partial y} = \phi_3 \tag{A.73}$$

$$\frac{\partial \phi_2}{\partial x} + \frac{\partial \phi_3}{\partial y} = f(x, y, \phi_1) \tag{A.74}$$

The functional J is taken to be

$$J = \int_{a_1}^{b_1} \int_{a_2}^{b_2} \left[L + P_1\left(\phi_2 - \frac{\partial \phi_1}{\partial x}\right) + P_2\left(\phi_3 - \frac{\partial \phi_1}{\partial y}\right) + P_3\left(f - \frac{\partial \phi_2}{\partial x} - \frac{\partial \phi_3}{\partial y}\right) \right] dx \, dy \tag{A.75}$$

Varying ϕ_1, ϕ_2, and ϕ_3, we have the adjoint equations

$$\frac{\partial P_1}{\partial x} + \frac{\partial P_2}{\partial y} = -\frac{\partial L}{\partial \phi_1} - P_3 \frac{\partial f}{\partial \phi_1} \tag{A.76}$$

$$\frac{\partial P_3}{\partial x} + P_2 = 0 \qquad\qquad (A.77)$$

$$\frac{\partial P_3}{\partial y} + P_1 = 0 \qquad\qquad (A.78)$$

We wish to determine the significance of P_1, P_2, and P_3. Let us write Equation A.75 in the general form

$$J = \int_{a_1}^{b_1} \int_{a_2}^{b_2} dx\, dy\, F(x, y, \phi_1, \phi_2, \phi_3, \phi_{1x}, \phi_{2x}, \phi_{3x}, \phi_{1y}, \phi_{2y}, \phi_{3y})$$

$$(A.79)$$

and take variations to obtain

$$J = \int_{a_1}^{b_1} \int_{a_2}^{b_2} dx\, dy\, (F_{\phi_1}\,\delta\phi_1 + F_{\phi_{1x}}\,\delta\phi_{1x} + F_{\phi_{1y}}\,\delta\phi_{1y} + F_{\phi_2}\,\delta\phi_2$$

$$+ F_{\phi_{2x}}\,\delta\phi_{2x} + F_{\phi_{2y}}\,\delta\phi_{2y} + F_{\phi_3}\,\delta\phi_3 + F_{\phi_{3x}}\,\delta\phi_{3x} + F_{\phi_{3y}}\,\delta\phi_{3y})$$

$$(A.80)$$

Note that

$$F_{\phi_{1x}}\,\delta\phi_{1x} = \frac{\partial}{\partial x}(F_{\phi_{1x}}\,\delta\phi_1) - \delta\phi_1\frac{\partial}{\partial x}(F_{\phi_{1x}}) \qquad (A.81)$$

$$F_{\phi_{1y}}\,\delta\phi_{1y} = \frac{\partial}{\partial y}(F_{\phi_{1y}}\,\delta\phi_1) - \delta\phi_1\frac{\partial}{\partial y}(F_{\phi_{1y}}) \qquad (A.82)$$

and that similar expressions hold in the cases of $\delta\phi_2$ and $\delta\phi_3$. Equation A.80 can, therefore, be rewritten in the forms

$$\delta J = \int_{a_1}^{b_1} \int_{a_2}^{b_2} \left[\delta\phi_1\left(F_{\phi_1} - \frac{\partial}{\partial x}F_{\phi_{1x}} - \frac{\partial}{\partial y}F_{\phi_{1y}}\right) + \delta\phi_2\left(F_{\phi_2} - \frac{\partial}{\partial x}F_{\phi_{2x}} - \frac{\partial}{\partial y}F_{\phi_{2y}}\right) \right.$$

$$+ \delta\phi_3\left(F_{\phi_3} - \frac{\partial}{\partial x}F_{\phi_{3x}} - \frac{\partial}{\partial y}F_{\phi_{3y}}\right) + \frac{\partial}{\partial x}\left(F_{\phi_{1x}}\,\delta\phi_1 + F_{\phi_{2x}}\,\delta\phi_2 + F_{\phi_{3x}}\,\delta\phi_3\right)$$

$$\left. + \frac{\partial}{\partial y}\left(F_{\phi_{1y}}\,\delta\phi_1 + F_{\phi_{2y}}\,\delta\phi_2 + F_{\phi_{3y}}\,\delta\phi_3\right) \right] \qquad (A.83)$$

Since the Euler equations must be satisfied, this reduces to

$$\delta J = \int_{a_1}^{b_1} \int_{a_2}^{b_2} dx\, dy \left[\frac{\partial}{\partial x} (F_{\phi_{1x}} \delta\phi_1 + F_{\phi_{2x}} \delta\phi_2 + F_{\phi_{3x}} \delta\phi_3) \right.$$

$$\left. + \frac{\partial}{\partial y} (F_{\phi_{1y}} \delta\phi_1 + F_{\phi_{2y}} \delta\phi_2 + F_{\phi_{3y}} \delta\phi_3) \right] \qquad (A.84)$$

Performing the permitted integrations, we have

$$\delta J = \int_{a_2}^{b_2} dy\, (F_{\phi_{1x}} \delta\phi_1 + F_{\phi_{2x}} \delta\phi_2 + F_{\phi_{3x}} \delta\phi_3) \Big|_{a_1}^{b_1}$$

$$+ \int_{a_1}^{b_1} dx\, (F_{\phi_{1y}} \delta\phi_1 + F_{\phi_{2y}} \delta\phi_2 + F_{\phi_{3y}} \delta\phi_3) \Big|_{a_2}^{b_2}$$

$$(A.85)$$

Suppose that the boundary conditions are satisfied at b_1 and b_2. Then

$$\delta J = \int_{a_2}^{b_2} dy\, (F_{\phi_{1x}} \delta\phi_1 + F_{\phi_{2x}} \delta\phi_2 + F_{\phi_{3x}} \delta\phi_3) \Big|_{x=a_1}$$

$$- \int_{a_1}^{b_1} dx\, (F_{\phi_{1y}} \delta\phi_1 + F_{\phi_{2y}} \delta\phi_2 + F_{\phi_{3y}} \delta\phi_3) \Big|_{y=a_2}$$

$$(A.86)$$

Note from Equation A.75

$$F_{\phi_{1x}} = -P_1 \qquad\qquad\qquad (A.87)$$

$$F_{\phi_{1y}} = -P_2 \qquad\qquad\qquad (A.88)$$

$$F_{\phi_{2x}} = -P_3 \qquad\qquad\qquad (A.89)$$

$$F_{\phi_{2y}} = 0 \qquad\qquad\qquad (A.90)$$

$$F_{\phi_{3x}} = 0 \tag{A.91}$$

$$F_{\phi_{3y}} = -P_3 \tag{A.92}$$

Equation A.86 thus becomes

$$\delta J = \int_{a_2}^{b_2} dy \, (P_1 \, \delta\phi_1 + P_3 \, \delta\phi_2)\Big|_{x=a_1} + \int_{a_1}^{b_1} dx \, (P_2 \, \delta\phi_1 + P_3 \, \delta\phi_3)\Big|_{y=a_2}$$

$$\tag{A.93}$$

The adjoint function may again be considered in terms of the effects of boundary conditions on the value of the functional, since

1. If $\delta\phi_1 = \delta(y - y_0)$, then $\delta J = P_1(a_1, y_0)$

2. If $\delta\phi_1 = \delta(x - x_0)$, then $\delta J = P_2(x_0, a_2)$

3. If $\delta\phi_2 = \delta(y - y_0)$, then $\delta J = P_3(a_1, y_0)$ (A.94)

4. If $\delta\phi_3 = \delta(x - x_0)$, then $\delta J = P_3(x_0, a_2)$

These boundary interpretations are easily extended off the boundaries.

In general, the ease of applying interpretation to the adjoint function in a multidimensional problem depends on the types of nonlinearities in the governing equation. An equation of the form

$$\frac{\partial^2\phi}{\partial x^2} + \phi\frac{\partial^2\phi}{\partial y^2} = f(\phi, x, y) \tag{A.95}$$

would be written in the canonical form

$$\frac{\partial\phi_1}{\partial x} = \phi_2 \tag{A.96}$$

$$\frac{\partial\phi_1}{\partial y} = \phi_3 \tag{A.97}$$

$$\frac{\partial\phi_2}{\partial x} + \phi_1\frac{\partial\phi_3}{\partial y} = f \tag{A.98}$$

When these equations are inserted in a functional of the type of Equation A.75, we would obtain $F_{\phi_{3y}} = -\phi_1 P_3$, and 4 in Equation A.94 would become

4.   If $\delta\phi_3 = \delta(x - x_0)$,   then $\delta J = P_3(x_0, a_2) \; \phi(x_0, a_2)$

Since ϕ would not be known a priori, this effect on J might be limited in its utility. On the other hand, 1, 2, and 3 could still be applied.

In the development following Equation A.84, it was assumed for simplicity that the boundaries were rectangular in the variables x, y (although it was not assumed that these coordinates were Cartesian). If the boundaries are of arbitrary shape, then Equation A.86 still holds with

$$a_1 = a_1(y) \tag{A.99}$$

$$a_2 = a_2(x)$$

Note of Equation A.99 must be taken in Equation A.94. In 1 of Equation A.94, for example, we would have

$$\delta J = P_1[a_1(y_0), y_0)] \tag{A.100}$$

This type of modification would be made in 2, 3, and 4 of Equation A.94 also.

In this section we have tried to illustrate how interpretation may be assigned in nonlinear problems with more than one independent variable, by treating a particular Equation A.68. Similar procedures may be applied to higher-dimensional and higher-order problems, and similar pitfalls may be encountered.

It is of interest to note that placing the equations in canonical forms, such as Equations A.36, A.72, A.73, and A.74, yields information not only about the scalar adjoint function but also about several of its derivatives. We consequently obtain the effect of perturbations not only in the system function but also in several of its derivatives. In a neutron-diffusion problem, for example, we obtain the effects of perturbations in flux and current.

Appendix B

ELEMENTS OF THE CALCULUS OF VARIATIONS

This appendix provides a review of background information on the calculus of variations, a knowledge of which is assumed in the main text. (For a more thorough treatment, see the work of Elsgolc.[6]) The requirement that a functional be stationary with respect to arbitrary variations in a function leads to an "Euler equation," and the boundary conditions of this equation can be modified by adding boundary terms to the functional. Types of functionals that lead to ordinary and to partial differential equations are noted. The Lagrange multiplier technique for treating problems with constraints is discussed.

B.1 The Basic Problem of the Calculus of Variations

The calculus of variations is concerned with finding functions that make functionals stationary. A functional is a variable that assumes a specific numerical value for each function which is substituted into it. A simple example is

$$J(y) = \int_a^b dx \, y(x) \tag{B.1}$$

Each function $y(x)$ yields a single numerical value of the functional J. For this reason, a functional is sometimes referred to as a function of a function.

Given a functional

$$J = J(y) \tag{B.2}$$

the fundamental problem of the calculus of variations is to find a function $y(x)$, such that increments $\delta y(x)$ in this function yield only second-order increments in the functional J, that is, when

$$y(x) \rightarrow y(x) + \delta y(x) \tag{B.3}$$

$$J \rightarrow J + O\left[\max_x \delta y(x)\right]^2 \tag{B.4}$$

103

Placing Equation B.3 into Equation B.2 and requiring Equation
B.4, we have an equation determining y(x). This equation is
called the Euler equation for the problem. In the following sections, Euler equations will be obtained for several representa-
tive problems.

B.2 A Simple Problem

The case where the functional is of the form

$$J = \int_a^b dx\ F(x, y, y')$$ (B.5)

where the prime denotes the derivative dy/dx, is one of the simplest problems of the calculus of variations. Allowing y(x) to
vary as in Equation B.3 yields, to first-order in δy,

$$\delta J = \int_a^b dx \left(\frac{\partial F}{\partial y} \delta y + \frac{\partial F}{\partial y'} \delta y'\right)$$ (B.6)

where

$$\delta y = \frac{d}{dx} (\delta y)$$ (B.7)

Integrating by parts, we have

$$\delta J = \int_a^b dx\ \delta y \left(\frac{\partial F}{\partial y} - \frac{d}{dx} \frac{\partial F}{\partial y'}\right) + \frac{\partial F}{\partial y'} \delta y \Big|_a^b$$ (B.8)

If the function y(x) is constrained to satisfy the boundary conditions

$$y(a) = y_a \quad y(b) = y_b$$ (B.9)

then $\delta y(a)$ and $\delta y(b)$ must be zero and the boundary terms of Equation B.8 vanish. If not, then y(x) must satisfy the "natural" boundary conditions

$$\frac{\partial F}{\partial y'} (a) = \frac{\partial F}{\partial y'} (b) = 0$$ (B.10)

Since $\delta y(x)$ is arbitrary (aside from possibly having to satisfy boundary conditions), $y(x)$ must satisfy the equation

$$\frac{\partial F}{\partial y} - \frac{d}{dx}\frac{\partial F}{\partial y'} = 0 \qquad (B.11)$$

if the first-order change in J is to be zero. Equation B.11, an ordinary differential equation, is called the Euler equation.

To determine whether this "stationary" value of J corresponds to a maximum, a minimum, or a saddle point, one may examine the behavior of the second variation $\delta^2 J = \delta(\delta J)$. If the second variation is positive definite, negative definite, or of indeterminate sign, then the value is a minimum, maximum, or a saddle point, respectively.

B.3 The Effect of Boundary Terms

In variational methods, we are often interested in obtaining equations with boundary conditions other than Equations B.9 or B.10. The functional may be modified by addition of boundary terms, thus leading to an Euler equation with different boundary conditions. Let the functional Equation B.5, for example, be modified to

$$J = \int_a^b dx\ F(x, y, y') - g_1[y(a)] + g_2[y(b)] \qquad (B.12)$$

Following the procedure of Section B.2 again leads to the Euler Equation B.11, but the natural boundary conditions are modified to

$$\frac{\partial F}{\partial y'}(a) + \frac{\partial g_1}{\partial y}(a) = 0 \qquad (B.13)$$

$$\frac{\partial F}{\partial y'}(b) + \frac{\partial g_2}{\partial y}(b) = 0 \qquad (B.14)$$

B.4 Problems With Several Dependent Variables

If the functional depends on several functions $y_1, y_2, \ldots, y_n$ in the form

$$J = \int_a^b dx\ F(x, y_1, y_1', y_2, y_2', \ldots, y_n, y_n') \qquad (B.15)$$

we obtain a set of simultaneous Euler equations

$$\frac{\partial F}{\partial y_i} - \frac{d}{dx} \frac{\partial F}{\partial y_2'} = 0 \qquad (B.16)$$

where $i = 1, 2, \cdots, n$, with the boundary terms

$$\frac{\partial F}{\partial y_i'} \delta y_i \bigg|_a^b = 0 \qquad (B.17)$$

where $i = 1, 2, \cdots, n$.

B.5 Problems With Several Independent Variables

If the function is defined over several dimensions in a functional of the form

$$J = \int_{a_1}^{b_1} dx_1 \cdots \int_{a_n}^{b_n} dx_n \; F(x_1, x_2, \cdots, x_n, \; y, y_{x1}, y_{x2}, \cdots, y_{xn}) \qquad (B.18)$$

where

$$y_{xi} = \frac{\partial y}{\partial x_i} \qquad i = 1, 2, \cdots, n \qquad (B.19)$$

then the Euler equation is the partial differential equation

$$\frac{\partial F}{\partial y} - \frac{\partial}{\partial x_1} \frac{\partial F}{\partial y_{x1}} - \frac{\partial}{\partial x_2} \frac{\partial F}{\partial y_{x2}} - \cdots - \frac{\partial}{\partial x_n} \frac{\partial F}{\partial y_{xn}} = 0 \qquad (B.20)$$

B.6 Functionals With Higher-Order Derivatives

If Equation B.5 is modified to

$$J = \int_a^b dx \; F(x, y, y^{(1)}, y^{(2)}, \cdots, y^{(n)}) \qquad (B.21)$$

where

$$y^{(i)} = \frac{d^i y}{dx^i} \tag{B.22}$$

then n integrations by parts will be required to obtain the Euler equation, which is

$$\frac{\partial F}{\partial y} - \frac{d}{dx} \frac{\partial F}{\partial y^{(1)}} + \cdots + (-1)^n \frac{d^n}{dx^n} \frac{\partial F}{\partial y^{(n)}} = 0 \tag{B.23}$$

This is, in general, an equation of order $2n$.

B.7 Problems With Constraints; Lagrange Multipliers

Suppose that Equation B.5 is to be made stationary subject to the constraint

$$\int_a^b dx\, G(x, y, y') = 0 \tag{B.24}$$

The constraint prohibits us from taking completely arbitrary variations in $y(x)$, so Equation B.11 is not the Euler equation for this problem.

The Lagrange multiplier approach is to multiply Equation B.24 by an arbitrary constant λ and add the resulting term to the functional, yielding

$$J = \int_a^b dx\, [F(x, y, y') + \lambda G(x, y, y')] \tag{B.25}$$

Varying y, we have

$$\delta J = \int_a^b dx\, \delta y \left[\frac{\partial}{\partial y}(F + \lambda G) - \frac{d}{dx} \frac{\partial}{\partial y'}(F + \lambda G) \right] \tag{B.26}$$

Since λ is an arbitrary parameter it may be <u>chosen</u> so that the equation

$$\frac{\partial}{\partial y}(F + \lambda G) - \frac{d}{dx} \frac{\partial}{\partial y'}(F + \lambda G) = 0 \tag{B.27}$$

is satisfied. We may solve Equation B.27 for $y(x, \lambda)$ and invoke Equation B.24 to solve for λ. The parameter λ chosen in this manner is called a Lagrange multiplier.

Another type of constraint that is often applied to the functional Equation B.5 is

$$V(x, y, y') = 0 \qquad\qquad (B.28)$$

The procedure is similar, except the multiplier is a function of x. The Euler equation becomes

$$\frac{\partial}{\partial y}\left[F + \lambda(x)\ V\right] - \frac{d}{dx}\left\{\frac{\partial}{\partial y'}\left[F + \lambda(x)\ V\right]\right\} = 0 \qquad (B.29)$$

Equations B.28 and B.29 are solved simultaneously for $y(x)$ and $\lambda(x)$.

Appendix C

ON OBTAINING AN APPROPRIATE FUNCTIONAL

In Section 3.1, a set of criteria is put forth that should be satisfied by variational methods. In this appendix it will be shown that the requirements of simplicity and an accurate value of a positive-definite quantity related to the residual implies a functional of the least-squares type.

A general functional that yields an accurate residual is

$$I = \int_a^b dx \, h(\phi, x)(H\phi - f) \tag{C.1}$$

The first variation is

$$\delta I = \int_a^b dx \, [\delta h(H\phi - f) + h\delta(H\phi)] \tag{C.2}$$

The definition of an adjoint operator H^* yields

$$\int_a^b dx \, h\delta(H\phi) = \int_a^b dx \, \delta\phi H^* h \tag{C.3}$$

The first term in Equation C.2 can be written in the form

$$\int_a^b dx \, \delta h(H\phi - f) = \int_a^b dx \, \delta\phi M(H\phi - f) \tag{C.4}$$

where the operator M is determined by the dependence of h on ϕ and on the derivatives of ϕ. Equation C.2 may therefore be written

$$\delta I = \int_a^b dx \, \delta\phi[M(H\phi - f) + H^* h] \tag{C.5}$$

and the Euler equation becomes

$$M(H\phi - f) + H^*h = 0 \tag{C.6}$$

which must be put in the form

$$G(H\phi - f) = 0 \tag{C.7}$$

The simplest general way to accomplish this is to make

$$h = g(\phi, x)(H\phi - f) \tag{C.8}$$

where g is an arbitrary function of ϕ and x. We then obtain

$$G = M + H^*(g \cdots) \tag{C.9}$$

and Equation C.1 becomes

$$I = \int_a^b dx\, g(\phi, x)(H\phi - f)^2 \tag{C.10}$$

If I is to be positive-definite, then so must $g(\phi, x)$.

Since g is arbitrary, it may be taken to be unity, and Equation C.10 becomes the simple least-squares functional

$$I = \int_a^b dx\, (H\phi - f)^2 \tag{C.11}$$

Equations C.3 and C.4 are then the same, so the Euler equation becomes

$$H^*(H\phi - f) = 0 \tag{C.12}$$

This development is sufficient to prove the assertion made in Section 3.1 that an operator G which makes Equation C.7 self-sufficient exists for all Equations 3.1. A positive weighting function p(x) may be included in h changing Equation C.1 into the more general form

$$I = \int_a^b dx\, p(x)(H\phi - f)^2 \tag{C.13}$$

The Euler equation then becomes

$$H^*[p(x)(H\phi - f)] = 0 \tag{C.14}$$

Appendix D

EIGENFUNCTIONS AND ERROR INTERPRETATIONS

In Section 3.7, use is made of the properties of the eigenfunctions of self-adjoint and non-self-adjoint homogeneous equations, and of the error interpretations of the Fourier expansions made with these eigenfunctions. This appendix summarizes these properties.

D.1 Orthogonal and Biorthogonal Modes

Consider the general linear homogeneous equation

$$H\psi_n = \lambda_n M\psi_n \qquad (D.1)$$

where the subscript n refers to the n-th eigenfunction or eigenvalue. Suppose that both H and M are self-adjoint operators. It is then true (by definition) that

$$\int_a^b dx\, \psi_m H\psi_n = \int_a^b dx\, \psi_n H\psi_m \qquad (D.2)$$

$$\int_a^b dx\, \psi_m M\psi_n = \int_a^b dx\, \psi_n M\psi_m \qquad (D.3)$$

Since

$$\int_a^b dx\, (\psi_m H\psi_n - \psi_n H\psi_m) = \lambda_n \int_a^b dx\, \psi_m M\psi_n - \lambda_m \int_a^b dx\, \psi_n M\psi_m \qquad (D.4)$$

it follows from Equations D.2 and D.3 that

$$(\lambda_n - \lambda_m) \int_a^b \psi_n M\psi_m\, dx = 0 \qquad (D.5)$$

111

This implies the orthogonality relation

$$\int_a^b \psi_n M \psi_m \, dx = \delta_{nm} \tag{D.6}$$

in which the normalization has been chosen arbitrarily.

Suppose, on the other hand, that H and M are not both self-adjoint. We may then consider the adjoint equation

$$H^* \psi_m^* = \lambda_m M^* \psi_m^* \tag{D.7}$$

By a procedure similar to the one above for self-adjoint equations, we obtain the biorthogonality conditions

$$\int_a^b dx \, \psi_m^* M \psi_n = \delta_{nm} \tag{D.8}$$

$$\int_a^b dx \, \psi_n M^* \psi_m^* = \delta_{nm} \tag{D.9}$$

It is thus seen that self-adjoint equations yield orthogonal (possibly with a weighting operator) eigenfunctions and non-self-adjoint equations yield biorthogonal eigenfunctions.

D.2 Minimum and Stationary Errors

Suppose we wished to expand a function $f(x)$ in a set ψ_n which is orthonormal with the weighting function $q(x)$

$$f(x) = \sum_{n=1}^N a_n \psi_n(x) \tag{D.10}$$

Suppose further that the coefficients a_n are to be chosen so as to minimize the weighted-square error

$$E = \int_a^b dx \left[f(x) - \sum_{n=1}^N a_n \psi_n \right]^2 q(x) \tag{D.11}$$

This objective implies the relations

$$\frac{\partial E}{\partial a_n} = 0 \quad n = 1, 2, \cdots, N \tag{D.12}$$

Applying Equation D.12, we have the Fourier coefficients

$$a_n = \int_a^b dx\ f(x)\ \psi_n(x)\ q(x) \tag{D.13}$$

If, on the other hand, the set ψ_n is biorthogonal with the set ψ_n^* with the weighting function $q(x)$, the expansion might also be made in the adjoint set

$$f(x) = \sum_{n=1}^N a_n^* \psi_n^*(x) \tag{D.14}$$

Suppose that the coefficients a_n and a_n^* are to be chosen so as to make stationary the weighted joint error

$$E = \int_a^b dx\left[f(x) - \sum_{n=1}^N a_n \psi_n\right]\left[f(x) - \sum_{n=1}^N a_n^* \psi_n^*\right] q(x) \tag{D.15}$$

This condition requires both Equation D.12 and

$$\frac{\partial E}{\partial a_n^*} = 0 \quad n = 1, 2, \cdots, N \tag{D.16}$$

The coefficients are then the Fourier coefficients

$$a_n = \int_a^b dx\ f(x)\ \psi_n^*(x)\ q(x) \tag{D.17}$$

$$a_n^* = \int_a^b dx\ f(x)\ \psi_n(x)\ q(x) \tag{D.18}$$

If the set ψ_n is orthogonal with a weighting operator M, then we might consider the expansion

$$f(x) = \sum_{n=1}^{N} b_n M \psi_n \tag{D.19}$$

and make stationary the joint error

$$E = \int_a^b dx \left[f(x) - \sum_{n=1}^{N} a_n \psi_n \right] \left[f(x) - \sum_{n=1}^{N} b_n M \psi_n \right] \tag{D.20}$$

yielding the coefficients

$$a_n = \int_a^b dx\; f(x)\; M\psi_n \tag{D.21}$$

$$b_n = \int_a^b dx\; f(x)\; \psi_n \tag{D.22}$$

In the biorthogonal case we might have

$$E = \int_a^b dx \left[f(x) - \sum_{n=1}^{N} a_n \psi_n \right] \left[f(x) - \sum_{n=1}^{N} b_n^* M^* \psi_n^* \right] \tag{D.23}$$

or

$$E = \int_a^b dx \left[f(x) - \sum_{n=1}^{N} b_n M \psi_n \right] \left[f(x) - \sum_{n=1}^{N} a_n^* \psi_n^* \right] \tag{D.24}$$

Making these errors stationary, we have

$$a_n = \int_a^b dx\; f(x)\; M^* \psi_n^* \tag{D.25}$$

$$b_n = \int_a^b dx\; f(x)\; \psi_n^* \tag{D.26}$$

$$a_n^* = \int_a^b dx \ f(x) \ M\psi_n \tag{D.27}$$

$$b_n^* = \int_a^b dx \ f(x) \ \psi_n \tag{D.28}$$

Referring to the eigenfunctions of Equation D.1, we may state the following conclusions:

1. If H is self-adjoint, and if M is a function, for example, M = q(x), then the Fourier coefficients for the expansion in eigenfunctions ψ_n have error-minimizing significance in the sense of Equation D.11.

2. If H and M are self-adjoint and if M is a noncommutative operator (that is, if $M\phi \neq \phi M$), or if either H or M is not self-adjoint, then the eigenfunction expansion has a stationary joint-error interpretation.

It should be noted that self-adjointness of an operator H - λM is not sufficient to guarantee an error-minimizing interpretation for the approximate representation, obtained when a function is expanded in the eigenfunctions of that operator.

REFERENCES

1. Bellman, R., and S. Dreyfus, Applied Dynamic Programming, Princeton University Press, Princeton, N. J. (1960), Chapter 5.

2. Bonilla, C., Nuclear Engineering, McGraw-Hill Book Co., Inc., New York (1955), p. 567.

3. Crandall, S., Engineering Analysis, McGraw-Hill Book Co., Inc., New York (1956).

4. Crandall, S., op. cit., pp. 286-287.

5. Dreyfus, S., RAND Corporation, Report No. P-2605 (1962).

6. Elsgolc, L., Calculus of Variations, Addison-Wesley Publishing Co., Inc., Reading, Massachusetts (1962).

7. Gelfand, I. M., and S. V. Fomin, Calculus of Variations, Prentice-Hall, Inc., Englewood Cliffs, N. J. (1963), Chapter 4.

8. Gelfand, I. M., and S. V. Formin, op. cit., Chapter 3.

9. Glasstone, S., and M. C. Edlund, The Elements of Nuclear Reactor Theory, D. van Nostrand Co., Inc., New York (1952), Chapter 6.

10. Hildebrand, F. B., Methods of Applied Mathematics, Prentice-Hall, Inc., Englewood Cliffs, N. J. (1952), Section 1.25.

11. Kaplan, I., "Nuclear Reactor Physics" (Printed Course Notes), M. I. T., Cambridge, Massachusetts, Chapter 19 (1960).

12. Kaplan, S., Nucl. Sci. and Eng., 13, 22 (1962).

13. Kaplan, S., O. Marlowe, and J. Bewick, Nucl. Sci. and Eng. 18, 163 (1964).

14. Lewins, J., Ph. D. Thesis, Department of Nuclear Engineering, M. I. T., Cambridge, Massachusetts (1959).

15. Lewins, J., J. Nucl. Energy, Part A, 13, 1 (1960).

16. Morse, P. M., and H. Feshbach, Methods of Theoretical Physics, McGraw-Hill Book Co., Inc., New York (1953), p. 1109.

17. Morse, P. M., and H. Feshbach, op. cit., pp. 870-874.

18. Morse, P. M., and H. Feshbach, op. cit., pp. 122-123.

19. Morse, P. M., and H. Feshbach, op. cit., Chapter 7.

20. Morse, P. M., and H. Feshbach, op. cit., pp. 877-878.

21. Nel, B., private communication (1946); Sc. D. Thesis (in preparation), Department of Nuclear Engineering, M.I.T., Cambridge, Massachusetts (1964).

22. Pomraning, G. C., and M. Clark, Jr., Nucl. Sci. Eng., 16, 147 (1963).

23. Poole, E., Introduction to the Theory of Linear Differential Equations, Dover Publications, Inc., New York (1960), p.1.

24. Reynolds, J., III, Sc. D. Thesis, Department of Mechanical Engineering, M.I.T., Cambridge, Massachusetts (1961).

25. Roussopolos, P., Compt. Rend., 236, 1858 (1953).

26. Selengut, D. S., AEC Report HW-59126, p. 89 (1959).

27. Selengut, D. S., op. cit., p. 103.

28. Todt, F., General Atomic, Report No. GA-2749 (1962).

29. Weinberg, A. M., and E. P. Wigner, The Physical Theory of Chain Reactions, University of Chicago Press, Chicago (1958), pp. 484-489.

30. Weinberg, A. M., and E. P. Wigner, op. cit., pp. 495-499.

INDEX